International Biotechnology Conference on Estate Crops 2017

IOP Conference Series: Earth and Environmental Science
Volume 183

Jakarta, Indonesia
18 - 20 October 2017

Editor:

Ir. Sumaryono

ISBN: 978-1-5108-7009-3
ISSN: 1755-1307

Printed from e-media with permission by:

Curran Associates, Inc.
57 Morehouse Lane
Red Hook, NY 12571

Printed by Curran Associates, Inc. (2018)

For permission requests, please contact the Institute of Physics
at the address below.

Institute of Physics
Dirac House, Temple Back
Bristol BS1 6BE UK

Phone: 44 1 17 929 7481
Fax: 44 1 17 920 0979

techtracking@iop.org

Additional copies of this publication are available from:

Curran Associates, Inc.
57 Morehouse Lane
Red Hook, NY 12571 USA
Phone: 845-758-0400
Fax: 845-758-2633
Email: curran@proceedings.com
Web: www.proceedings.com

Table of contents

Volume 183

INTERNATIONAL BIOTECHNOLOGY CONFERENCE ON ESTATE CROPS 2017

18–20 October 2017, Jakarta, Indonesia

Accepted papers received: 1 August 2018
Published online: 20 August 2018

Preface

International Biotechnology Conference on Estate Crops 2017

Editorial board

Organising committee

Peer review statement

Papers

Molecular oil palm breeding at Sampoerna Agro

Enrique Ritter and D Asmono.....1

Applications and challenges of biotechnology in oil palm breeding

Aik Chin Soh.....5

Transformation of oil palm calli using CRISPR/Cas9 System: toward genome editing of oil palm

Asmini Budiani, R A Putranto, I Riyadi, Sumaryono, H Minarsih and R Faizah.....11

Semi-quantitative RT-PCR analysis of transcripts encoding protease inhibitor in *Hevea brasiliensis* Muell. Arg latex

Irfan Martiansyah, D M Amanah and R A Putranto.....18

Genetic similarity analysis of *Vanda tricolor* Lindley var. Suavis orchids using OPU 03 and OPU 16 RAPD marker

Tony Haryo Wibisono, Suratsih, E Yulianti, L Sugiyarto and I S Mercuriani.....27

Microscopic callus selection of sengon tree (*Falcataria moluccana*) putative tolerant to *Uromycladium falcatarium*

Asri Insiana Putri.....31

Somatic embryogenesis of sago palm (*Metroxylon sagu* Rottb.) through suspension culture technique

Imron Riyadi, D Efendi, B S Purwoko and D Santoso.....37

Somatic embryogenesis of sago palm (*Metroxylon sagu* Rottb.) from different origins in Indonesia

Masna Maya Sinta, I Riyadi and Sumaryono.....46

Embryogenic callus initiation from leaf explants of *Elaeis oleifera* x *Elaeis guineensis* (OxG) hybrids

Sumaryono, I Riyadi, R T Saptari, H Y Rahmadi and Ernayunita.....52

Optimization of simultaneous saccharification and fermentation in bioethanol production from sugarcane bagasse hydrolyse by *Saccharomyces cerevisiae* BTCC 3 using response surface methodology

Ahmad Thontowi, U Perwitasari, L N Kholida, Fahrurrozi, Yopi and B Prasetya.....58

Bio-organo-chemical fertilizers: a new prospecting technology for improving fertilizer use efficiency (FUE)

Didiek Hadjar Goenadi, A B Mustafa and L P Santi.....66

Biodegradability of oil palm cellulose-based bioplastics

Isroi, A Rahman and K Syamsu.....77

Introduction of bioprospecting opportunities for Indonesian mangrove species

Julkipli, R R Batubara, G E Jogia, I Batubara, K A Audah and K N Nunuk.....82

Effect of bio-silica on drought tolerance in plants

Laksmita Prima Santi, Nurhaimi-Haris and D Mulyanto.....87

"CIRAGI"- The microbial fermentation starter for developing excellent coffee flavour

Tri Panji, Priyono, Suharyanto and A Zainuddin.....95

The effect of biostimulant in root and population of phosphate solubilizing bacteria: a study case in upland rice

Agustin Sri Mulyatni, R H Praptana and D Santoso.....100

Plant biostimulant to improve crops productivity and planters profit

Djoko Santoso, A Gunawan, A Budiani, D A Sari and Priyono.....105

Sucrose analysis in leaf tissue of two contrasted rubber tree clones under abiotic and biotic stresses

Irfan Martiansyah, R A Putranto and N Khumaida.....111

Chemical and biological characteristics of selected wet soil in oil palm plantation attack with *Ganoderma* sp.

Happy Widiastuti, D Taniwiryono, H Hendarjanti, S Harjotedjo and M Hadi Sugeng W.....121

Modified *Ganoderma* selective medium to meet Indonesia's government regulation

Wiwit Iman Amanda and H T Prakoso.....128

The potency of an automated detection system for real-time pest monitoring from smartphones and spatial analysis for leaf-eating caterpillar attack evaluation

Henny Hendarjanti, S W Indratno, U T Ismunandar and M H Sugeng W.....134

Effectiveness of Weed Solut-Ion as herbicide adjuvant to control weeds in oil palm plantations

Prabawati Hyunita Putri and D Guntoro.....141

Inhibitory effect of phenolic acid on *Ganoderma boninense* enzyme as an approach on *Ganoderma* infection

Irma Kresnawaty, D D Eris, A S Mulyatni and H T Prakoso.....149

IOP Conf. Series: Earth and Environmental Science **183** (2018) 011001 doi:10.1088/1755-1315/183/1/011001

Preface

Biotechnology techniques are carrying forward plantation crop development in attempts to understand the genetic basis of the crops, to produce tissue-cultured plantlets on a large scale, to employ molecular breeding of new cultivars for increased yield, resistance to pests and tolerance to environmental stresses, and to use microorganisms for bio-based products, soil healthiness and waste treatment.

The Indonesian Research Institute for Biotechnology and Bioindustry (IRIBB) has organized an International Biotechnology Conference on Estate Crops 2017 on 18- 20 October 2017 in Grand Sahid Jaya Hotel Jakarta, Indonesia. This Conference was in conjunction with the World Plantation Conferences and Exhibition 2017 (WPLACE-2017) on 18 October 2017. WPLACE-2017 provided General Topics (GTs) on related subjects.

It is truly meaningful for the Indonesian Research Institute for Biotechnology and Bioindustry (IRIBB) to hold the first International Biotechnology Conference on Estate Crops 2017 (IBCEC-2017), an international forum for biotechnology and exposition. The institution was founded in 1901 by the Dutch colonial government under the name of Centrale Proefstations Vereniging (CPV). In 1957, during the expropriation firms of the Dutch plantations in Indonesia, CPV was repossessed by the Indonesian government. For many years since the CPV, it has undergone many name changes until 2015 in which the Indonesian Research Institute for Biotechnology and Bioindustry (IRIBB) was officialised. The IRIBB carries a vision to be an innovative and internationally research-based institution in biotechnology and bioindustry.

The conference program included honourable lectures from eminent experts on the success stories and the recent development of biotechnology such as bioinformatics, genome editing, genomic selection, *in vitro* propagation, phytobiome, biostimulant, bioplastic, biopesticide, and biofermentation.

IOP Conf. Series: Earth and Environmental Science **183** (2018) 011002 doi:10.1088/1755-1315/183/1/011002

Editorial board

IOP Conf. Series: Earth and Environmental Science **183** (2018) 011003 doi:10.1088/1755-1315/183/1/011003

Organising committee

Advisory board	: Dr. Ir. Priyono, DIRS
	Dr. Asmini Budiani, MSi
	Dr. Laksmita Prima Santi, MSi
	Dr. Nurhaimi Haris, MSi
Chairperson	: Dr. Asmini Budiani, MSi
Co-Chairperson	: Ir. Sumaryono, MSc
Executive secretary	: Dr. Riza Arief Putranto, DEA
Treasury	: Tresna Wardana, SE
Secretarial	: Firda Dimawarnita, ST
	Dini Astika Sari, MBiotech
	Herti Sugiarti, AMd
	Fajar Prayoga, SKom
	Fenti Rachmawanty, AMd
	Ria Rahmatika, SE
Program and Proceeding	: Dr. Hayati Minarsih, MSc
	Irma Kresnawaty, MSi
	Irfan Martiansyah, MSi
	Haryo Tejo Prakoso, MAgrEnv
	Dian Mutiara Amanah, SSi
	Ciptadi Achmad Yusup, MSi
Sponsorship	: Dr. Happy Widiastuti, MSi
	Dr. Laksmita Prima Santi, MSi
	Dr. Hayati Minarsih, MSc
	Dr. Imron Riyadi, MSi
	Dr. Isroi, MSi
	Irma Kresnawaty, MSi

Peer review statement

All papers published in this volume of *IOP Conference Series: Earth and Environmental Science* have been peer reviewed through processes administered by the proceedings Editors. Reviews were conducted by expert referees to the professional and scientific standards expected of a proceedings journal published by IOP Publishing.

IOP Conf. Series: Earth and Environmental Science **183** (2018) 012001 doi:10.1088/1755-1315/183/1/012001

Molecular oil palm breeding at Sampoerna Agro

Enrique Ritter[1]* and D Asmono[2]

[1]NEIKER-Tecnalia, Vitoria-Gasteiz, Spain
[2]Sampoerna Agro Tbk PT, Palembang, Indonesia

*E-mail: eritter@neiker.eus

Abstract. Sampoerna Agro and NEIKER perform jointly since many years collaborative R&D projects related to Molecular breeding and Marker assisted selection in Oil palm in order to accelerate the oil palm breeding program and to improve its efficiency. The detection of candidate genes for useful traits offers the possibility to apply them – after developing the corresponding markers – in marker assisted selection (MAS) within breeding programmes. The survey of the allelic diversity of such genes in available germplasm and the analyses of their particular effects permits to select the most efficient alleles or allele combinations for these purposes. Within this project, we have identified useful candidate genes for different productive and quality traits as well as for stress tolerance using different molecular tools. We have characterized the allelic variation in this germplasm in order to develop marker assisted breeding strategies through Model building. The developed Models which show large correlations between predicted and observed values explaining large portions of the total variance can be used to assign breeding values to parents and to predict progeny performances of their crosses. In this paper, we summarize the R&D activities which are being performed and practically implemented at Sampoerna.

1. Introduction

Sampoerna Agro and NEIKER perform jointly since many years collaborative R&D projects related to Molecular breeding and Marker assisted selection in Oil palm in order to accelerate the oil palm breeding program and to improve its efficiency. Molecular breeding can accelerate considerably breeding programmes, particularly in oil palm with long vegetative cycles and extended selection periods. The detection of candidate genes for useful traits offers the possibility to apply them – after developing the corresponding markers – in marker assisted selection (MAS) within breeding programmes [1]. The survey of the allelic diversity of such genes in available germplasm and the analyses of their particular effects permits to select the most efficient alleles or allele combinations for these purposes. Within this project we have identified useful candidate genes for different productive and quality traits as well as for stress tolerance using different molecular tools; we have characterized the allelic variation in this germplasm in order to develop marker assisted breeding strategies through Model building [2]. The developed Models which show large correlations between predicted and observed values explaining large portions of the total variance can be used to assign breeding values to parents and to predict progeny performances of their crosses. In this paper, we summarize the R&D activities which are being performed, as well as their concrete, practical implementation at Sampoerna.

IOP Conf. Series: Earth and Environmental Science **183** (2018) 012001 doi:10.1088/1755-1315/183/1/012001

Useful traits of interest (such as yield, oil production, oil quality and tolerances to biotic and abiotic stresses) are controlled by the effects of one, several or many different, so called Candidate Genes (CG) in each case. Their numbers depend on the particular trait. The particular phenotype of an individual depends on the genotype and is controlled by the effects of the specific alleles of the candidate genes of the corresponding trait and possible interactions [3]. In order to analyse the effects of the different, existing alleles of candidate genes it is necessary to link phenotypic and genotypic data, applying statistical methods such as Association Mapping.

2. Materials and Methods

2.1 Plant material

a) Parental lines for producing *Tenera* seeds through DxP crosses belonging to different *Pisifera* and *Dura* origins.
b) 440 DxP progenies descending from selected palms from above, which have been extensively characterized (over 15 years) with respect to production traits, bunch component traits and other traits of interest. 220 selected genotypes of these progenies with the positive and negative expression for traits of interest have been used for molecular characterization.
c) New germplasm accessions from recent prospections in Nigeria, Angola and Cameroon.
d) Accessions from Field Trials for evaluating drought resistance and nutrient use efficiency.

2.2 Candidate gene (CG) detection different tools

- Bulked Segregant Analyses coupled to differential cDNA-AFLP Analyses which allow the direct detection of potential CG.
- Co-location Analyses of sequenced markers, QTL and mapped genes of a high-density oil palm reference map [4].
- Known Candidate genes from oil palm or other species or key enzymes from relevant metabolic pathways (Palm Oil Biosynthesis, Transcription factors) using in silico mining or other Bioinformatics approaches.

2.3 Molecular analyses

For the molecular analyses, Primers were designed for each CG to amplify CG exon sequences using Primer 3 Software and validated first on a small set of genotypes.

Each CG was first amplified with a fusion primer pair containing a CG specific part and a universal part. Subsequently, each sample was re-amplified with a so called Key Primer composed of a part specific to the sequencing machine, a "Barcode" (=MIDS or Index) for each genotype and the universal part of the fusion primers. Barcodes and CG specific parts identify the origin of the sample. Samples were pooled and sent for sequencing using ION torrent, ION proton or MySeq Illumina platforms.

For Data analyses, in house developed Software was used for sequence processing, analyses of the allelic variability of CG in the collection, determination of allele composition of the evaluated genotypes and determination of effects of CG alleles and allele combinations through association mapping (AM). In addition, whole genome association studies (WGAS) and genomic selection (GS) methods are applied to the available data.

For special genes of interest, such as the Shell (*Sh*) gene or the Virescence (*Vir*) Gene, allele-specific primers were designed for routine selection of desired plant materials.

3. Results and Discussion

Hundreds of CG were identified, processed and analysed as described above. Numerous significant effects of alleles and allele combinations (AC) were detected for different CG for all traits of interest.

Beside the analyses of individual CG, also their cumulative effect was analysed by model building. Based on average allele and AC values, trait specific breeding values of the parental lines were determined based on the allele composition of the parents for each CG. Moreover, progeny performance predictions were computed based on average parental breeding values or expected AC values for all CG. Highly significant correlations between the predicted and observed progeny performances were detected always.

The general applicability of the concept and the developed models was validated using subsets of parental data and even external data predicted with the allele or AC values of the original models. Based on these results, predicted progeny performance (PPP) matrixes for all possible crosses between available Dura and Pisifera parents were computed for each trait. These analyses were extended to combined predictions for multiple traits based on breeder preferences. Crosses were ranked based on these final scores and the resulting TOP crosses were extracted (**Figure 1**).

Based on these predictions, three comparative classical and molecular breeding field trials have been established targeting different combinations of traits of interest and involving a total of 48 crosses. They will be evaluated in the near future.

Ranking Values based on sum of Finals Score - Model: xxxx PCLAd q for Traits: FFB, BN, HI and OP

DU|		P1	P2	P3	P4	P5	P6	P7	P8	P9	P10	P11	P12	P13	P14	P15	P16	P17	P18	P19	P20	P21	P22	P23	P24	P25	P26	P27	P28	P29	P30	P31	P32	P33	P34	P35	P36	P37	P38	P39	P40	P41	P42	P43	P44	P45	P46	P47	P48
	Nar		P3(	P2/	P3/	P4/	P5/	P30	P36	P8/	P9/	P10	P11	P30	P30	P14	P15	P16	P17	P36	P35	P20	P36	P22	P23	P24	P25	P26	P36	P28	P29	P30	P31	P32	P33	P34	P35	P36	P37	P32	P39	P40	P41	P42	P43	P35	P36	P46	P47	P48
D1	D1						D												D		D			C	C	C		D	B				C	C	D	C	C	D			D		D							C	C	
D2	D2																				D								D					D				D												D		
D3	D3						D												D		D			D	D	D			C				D	C		D	D	D					D								C	D	
D4	D4																																																	
D5	D5																																																	
D6	D6																				D								D					D				D												D		
D7	D7																							C	D	C		D	C				C	D	D	C	D														D	C	
D8	D8																											D																							
D9	D9		D				D				D						D						D	D	C	D	D	B	C	C		C	A				B	B	C	B	C	C			D		D			D		D		B	B	D
D10	D10																																																	
D11	D11																				D			B	D	C		C	B				B	C	C	B	D	D									D		C	B	
D12	D12																							C	D	C		D	C				C	D	D	C	D														D	C	
D13	D13																																																	
D14	D14						D												D		D			C	C	C		D	B				C	C	D	C	C	D			D		D							C	C	
D15	D15																				D			D				D	C				D	D		D		D													D	D	
D16	D16																																																	
D17	D17																							C	D	C		D	C				C	D	D	C	D														D	C	
D18	D18																							C	D	C		D	C				C	D	D	C	D														D	C	
D19	D19						D												D		D			C	D	C		C	C				C	D	D	C	D	D													D	C	
D20	D20																																																	
D21	D21																							D		D		D	D				D			D																	D	
D22	D22																							C		D		D	C				C	D	D	C																D	C	
D23	D23																							D		D			D				D	D		D																D	D	
D24	D24																																																	
D25	D25																							C	D	C		D	C				C	D	D	C	D														D	C	
D26	D26																																																	
D27	D27																							D		D		D	D				D			D																	D	
D28	D28																											D																							
D29	D29						D												D		D			D		D		D	D				D	D		D		D													D	D	
D30	D30																							D					D				D			D																	D	
D31	D31																							C	D	C		D	C				C	D	D	C	D														D	C	
D32	D32																																																	
D33	D33																							C	D	C		D	C				C	D	D	C	D														D	C	
D34	D34																				D								D					D				D												D		
D35	D35		C				C				C						C	C	B		C	C	D	D		C	B				C	B	D	C	D	B					D							B	C	

Figure 1. Visualization of TOP crosses for combined trait scores (A=12 points, B, C, D = 11,10 and 9 points, respectively).

Another example for the practical implementation of Molecular Breeding at Sampoerna is the routine screening of plant material from the breeding program with allele specific primers for the *SH* gene and *Vir* gene, respectively. For 2017 the analyses of a total of 1800 samples are targeted.

Pisifera parents are obtained from TxP or TxT crosses. The identification of the fruit type already in seedling stage allows to select and plant only the desired Pisifera genotypes and to save 50 or 75% of land for cultivation. A similar example represents the DxT crosses to evaluate the performance of Tenera genotypes.

The Vir allele screening for developing virescence cultivars allows detecting even the desired homozygous genotypes from crosses between heterozygous parents, which using only classical breeding would require one additional generation of selfings.

For preselecting of new germplasm from Nigeria, Angola or Cameroon, also simple molecular markers such as SSR are applied to select appropriate subsets of diverse materials based on the results of Cluster analyses and resulting dendrograms [2] (**Figure 2**).

IOP Conf. Series: Earth and Environmental Science 183 (2018) 012001 doi:10.1088/1755-1315/183/1/012001

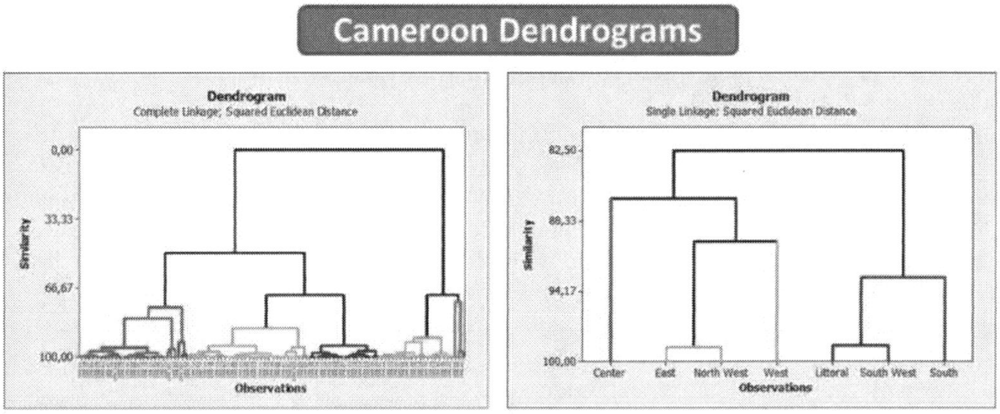

Figure 2. Example of pre-selection of plant materials based on Cluster analyses derived from SSR markers.

Moreover, *Ganoderma* Tolerance Screenings are performed in the breeding material using promising candidate genes and the new traits of interest are targeted such as the ability for High Density Planting (HDP) and Sex Ratio.

4. Conclusions

In classical breeding, the breeding value of a parent is deduced from the progeny performances he is involved. Unfortunately, in oil palm, only a reduced number of progenies are available to evaluate a parent (2-3 crosses for Dura, 6-8 crosses for Pisifera parents).

However, many parents share the same allele of a particular CG. Parental alleles of CG are known and the information of all available progenies can be used to calculate the effects of CG alleles or allele combinations, providing more information in this way.

Molecular breeding allows to select already at the seedling stage the genotypes of interest and to plant only the desired plant material. The next Breeding Cycle can be already I initiated at flowering time and it is also possible to plant only the most promising Crosses based on improved progeny performance predictions.

Finally, by inspecting carefully the allelic composition of genotypes, it is possible to pyramidize favourable alleles of CG in the parents and to apply within selection in progenies.

5. References

[1] Crossa J, de los Campos G, Perez P, Gianola D, Burgueño J, Araus J L, Makumbi D, Singh R P, Dreisigacker S, Jianbing Yan, Arief V, Banziger M and H-J Braun 2010 *Genetics* **186** 713

[2] Ritter E, Pratiwi E, Lopez de Armentia E, Herrero J, Ningrum P Y, Santika B, Yulismawati, Sarimana U, Sembiring Z, Asmono D and Hernandez M *Euphytica* **2** 207

[3] Durstewitz G, Polley A, Plieske J, Luerssen H, Graner EM, Wieseke R and Ganal M W 2010 *Genome* **53** 948

[4] Wang N, Chen B, Xu K, Gao G, Li F, Qiao J, Yan G, Li J, Li H and Wu X 2016 *Front Plant Sci.* **24** 338

IOP Conf. Series: Earth and Environmental Science **183** (2018) 012002 doi:10.1088/1755-1315/183/1/012002

Applications and challenges of biotechnology in oil palm breeding

Aik Chin Soh[1*]

[1]Crops For the Future and School of Biosciences, University of Nottingham Malaysia Campus, Semenyih, Selangor, Malaysia

*E-mail: sohac28@gmail.com

Abstract. Biotechnological applications in oil palm breeding comprises tissue culture propagation (TC), genetic modification (GM), marker-assisted breeding (MAB), and marker-assisted selection (MAS). Commercial success of oil palm TC has been hampered by the fruit mantling somaclonal variation risk, inefficient cloning process and inefficient ortet selection. With the imminent availability of the mantling marker and subsequently markers for cloning and palm selection efficiencies, the scenario would change. Despite early enthusiasm, GM oil palm remains at the R&D level and only for the high oleic trait, because of technical complexity and public sensitivity. Marker-assisted breeding, population genetic diversity, homozygosity determination and linkage disequilibrium have made most impact in oil palm breeding. With the revelation of the oil palm whole genome sequence, the feasibility of MAS for many major quantitative gene locus (QTL) traits e.g. shell, fruit colour, mantling, long stalk, lipase are being expedited. Meanwhile for minor QTL controlling yield and other traits, oil palm breeders are testing the feasibilities of genome-wide breeding/genetic value estimation for MAS of hybrid parents, families and individuals. Undoubtedly, biotechnological tools will continue improving but their success depends on the field breeder who provides the essential genetic trials to develop the biotechnological tools and to validate their utility.

1. Introduction

The oil palm is a relatively new crop with about a hundred years' history [1]. With a generation cycle of about 8-10 years, it has undergone only about 10 cycles of scientific plant breeding. Nevertheless, through breeding cum agronomy research, yields of less than 1 t/ha oil in the wild palm groves have been improved to 7-9 t/ha achievable on a plantation scale to 18 t/ha potential yield. Oil palm breeding has also achieved 1-1.5% yield improvement per year comparable to those achieved by major crops [2].

The main objective of oil palm breeding since the early days has been high and precocious oil yield. This has resulted in the development of vigorous tall growing varieties that poses increasing manual harvesting problems in labour scarce plantation areas. The extensive and intensive (repeated) plantings of uniform narrow genetically based bred varieties responsive to high agronomic inputs and good growing

IOP Conf. Series: Earth and Environmental Science **183** (2018) 012002 doi:10.1088/1755-1315/183/1/012002

conditions have also led to their genetically vulnerability to existing and new pest and diseases (P&D): leaf-eating insect pest attacks e.g. leaf miners; root, stem and bud rots e.g. Ganoderma Basal Stem Rot, Fusarium Wilt, Lethal Bud Rot and to the acute environmental stresses (global warming, extreme temperature and rainfall fluctuations, storms) attributed to climate change arising from anthropogenic CO_2 rise in the atmosphere. The oil palm has to adapt (via resistance/tolerance/resilience) to these stresses. It also has to mitigate against CO_2 and greenhouse gas rise by being more efficient in its use of resource inputs e.g. N fertilizer, water, and light. Hence new breeding objectives and approaches are needed: breeding for resource use efficiency (RUE); breeding for resistance/tolerance/resilience to biotic (P&D) and abiotic (environmental) stress, and breeding for biodiversity.

Recognizing that the traditional breeding approach requires 20-40 years to produce a new variety (depending on the starting genetic base) will grossly be inadequate and slow to respond to the Industry needs to meet the rapidly changing economic and environmental challenges, oil palm breeders have turned to biotechnology as by other crop breeders.

2. Oil Palm Biotechnology

Biotechnology in oil palm breeding encompasses *in vitro* (tissue culture) propagation, genetic modification (GM) or transgenics, genome sequencing and molecular breeding (DNA marker-assisted breeding or MAB, and marker-assisted selection or MAS [3].

3. *In Vitro* Propagation

Oil palm tissue culture (OPTC) clonal propagation of elite palms was the first attempt into biotechnology [4, 5]. The rationale was that commercial oil palm hybrid progenies were genetically heterogeneous (arising from the use of non-fully inbred parents) with individual palms capable of yielding more than 30% of the mean. OPTC R&D began in the mid-1970s but only became a technology with available commercial plantlet sale in the new millennium [6]. However, to date clonal palms have yet to make a significant impact on field plantings constituting less than 2% of the oil palm seeds market and plantings [7]. The mantled fruit somaclonal variation risk [8] leading to palm sterility, cloning (callogenesis, somatic embryogenesis) inefficiency resulting in high production cost, and lower than expected yield advantage of clonal palms have been the attributing factors [6, 9, 10]. Recloning and liquid cultures would reduce the second and third inefficiencies but with the attendant higher risk of fruit mantling. The lag time with recloning would also reduce its advantage over concurrently improved hybrids.

Commercial production has been achieved through cloning a basket of ortets or clonal parents, limiting production per culture line from the gel and liquid suspension systems and from primary and secondary (reclones) clones to hedge on mantling risk and yield advantage yet achieving commercial production levels [6]. Packaging a mix of clones is also done in the field to hedge against mantling, extended juvenility, poor pollination, P&D and GxE (genotype x environment interaction) risks with the consequence of compromising on the maximum yield achievable with clones.

The molecular bases of fruit mantling and callogenesis cum somatic embryogenesis in oil palm clones have been actively researched [11, 12, 13]. Fruit mantling is the consequence of hypomethylation of the Karma transposon in the homeotic genes [14]. Histone modification may also be involved. Some prospective molecular markers for callogenesis and somatic embryogenesis are thus anxiously anticipated. When available, clonal propagation would expedite the production and planting of clonal palms not only for high oil yield but also for ease of harvesting, value addition traits (oil quality, vitamins, phytonutrients), P&D and environmental stress resistance/adaptation traits and also the early exploitation of new genetic material to reduce genetic vulnerability and enhance sustainability of the plantation [15].

4. Genetic Modification

Genetically modified (GM) oil palm has been researched since the 1990's. The primary traits targeted were high oleic acid for healthy low unsaturated oil [18, 19] and high PHBV (polyhydroxy butyrate valeate) for bioplastics production [18]. Such putative transformed palms have been obtained via biolistics and *Agrobacterium*-mediated methods [19, 20, 21, 22, 23] and are being validated for stable single gene copy incorporation and regeneration in the laboratory and biosafety containment houses. In common with the world trend, GM oil palm work has been downsized to just the oleic R&D program as proof of concept due to technical difficulties (multiple biochemical pathways, tissue-specific promotors and genes involved in many of the traits [24], cumbersome and biosafety test requirements and public sensitivity. Recently, the gene editing technique which allows targeted in situ DNA alteration without or transient use of transgenes has been touted as a solution to the controversy [25]. If so, GM oil palm research will resume its importance [24] especially with the revelation of the oil palm whole genome and some key trait sequence [26, 27, 28].

5. Marker-Assisted Breeding

5.1. DNA marker or molecular marker-assisted breeding (MAB)

MAB has made most impact in oil palm breeding: genotype authentication of breeding parents and hybrids, clonal ortets and ramets is now standard practice [29]. Besides facilitating accuracy and precision of breeding and cloning, it has application in plant variety protection (PVP) although currently it plays only as supporting evidence to phenotypic discrimination [30]. Population molecular diversity studies for breeding heterotic population organization, germplasm prospection and core collection which facilitate the selection, maintenance and exploitation of key genotypes and populations have also now become standard practice in many oil palm breeding programs [31, 32, 33, 34]. Related to these are also homozygosity testing which expedites development of inbred (also isogenic and recombinant inbred lines) via single seed descent or doubled haploidy [15, 35].

5.2. Genetic mapping, linkage studies and gene discovery

These serve as prelude to developing markers for key genes besides understanding their organization and function. This active area of research has culminated in the elucidation of three genome sequence maps for *E. guineensis* (MPOB, ACGT, Sime Darby) and one for *E. oleifera* by MPOB [26]. This has facilitated the discoveries of the genomic bases of the shell, virescens, and fruit mantling traits [14, 27, 28]. It is now apparent the oil palm genome varies with the different populations and likewise the genes controlling the traits e.g. shell, virescens with five separate mutations each found to date. As such, different or a suit of markers may need to be deployed in marker-assisted selection (MAS).

6. Marker-Assisted Selection

Being an outbred perennial tree crop, the oil palm lacks traditional genetic maps for phenotypic traits. With the progressive advent of fine or more precise DNA markers e.g. RFLP (randomized fragment length polymorphism, Amplified FLP, SSR (simple sequence repeats), SNP (single nucleotide polymorphism), culminating in the publication of the oil palm whole genome, and together with the reduced cost of high-throughput re-sequencing provided a better opportunity for high density precise mapping of important traits [19, 36].

MAS is currently actively pursued in oil palm breeding. MAS allows early and efficient selection saving time, space, effort and consequently money. Precise markers for the shell and virescens fruit traits are available; the former useful in separating out the tenera (T) palms for their dedicated evaluation from the complicating co-segregant dura (D) and pisifera (P) palms. The D palms are usually of little interest in such trials, while the Ps are of interest in DxP progeny-testing cum pollen source for seed production and preferably planted as a separate "pollen garden". The inclusion of the vigorous growing female sterile P segregants in various proportions in the trial TxT plots will also bias the agronomic evaluation of the T palms and plots. The shell marker is also useful in assuring the legitimacy of commercial pedigreed DxP seeds [37] although pedigree DNA fingerprinting using SNP/SSR would be better. Markers for other traits e.g. ease of harvesting (dwarf, long stalk, virescence, delayed fruit shedding, low lipase, high callogenesis/ embryogenesis are currently being sought and developed [23, 38, 39].

Traditional MAS deals with simple traits controlled by one or few major genes and the backcross breeding method is the most common approach and in biparental crosses. Quantitative traits e.g. yield, disease resistance, are controlled by quantitative trait loci (QTL) which may be a few with major effects, many with minor effects with or a combination of both and may be spread all over the genome rather than confined to certain chromosomes. This is particularly important for a population improvement (recurrent selection) breeding program.

Unlike linkage analysis, GWAS (genome-wide association studies) provides access to the total meiotic recombination accumulated in a large population during evolution to increase the mapping resolution. The GWAS-GEBV (genetic expected breeding value) uses thousands of SNPs to associate with the individual phenotypic differences in an essentially multiple regression approach. Only major QTL i.e. significant SNP-trait associations are used and may lose out on minor QTL with this approach [40]. Sub-population structure and kinship effects have to be removed. Significance testing level also needs to be very stringent ($\alpha = 10^{-5}$) to avoid false positive results. The method also focuses on common variation, many of the associated variants are not causal.

The GS-BLUP (genomic selection best linear unbiased prediction) or mixed model approach uses all the markers irrespective of statistical significance and is more favoured by plant and animal breeders currently [41, 42]. As in GWAS, a large population of individuals are genotyped cum phenotyped. All marker and phenotypic data are fitted into a single statistical (mixed) model, called model training. The model is then used to predict the breeding values of 'future' individuals that have genotyped but not phenotyped.

Applications of GWAS and GS in oil palm both for parent breeding value (BV) and hybrid family and individual palm genotypic value (GV) predictions are currently being field validated e.g AAR. Sime Darby [43, 44]. If proven successful, much breeding time and effort could be saved and expected breeding progress achieved. The BV approach being based on family mean data would likely to be the more useful approach. Individual plant data and selection is prone to environmental and genotype x environment interaction effect complications. Sample size, population structure and kinship issues would also need to be accounted for to obtain valid predictions. The GS approach also allows the incorporation of 'omics' (proteomics, metabolomics, phenomics) data which would more pertinent in breeding for disease resistance (host-parasite interaction) and stress adaptation (genotype x environment interaction) traits [45, 46, 47, 48].

7. Conclusions

Biotechnological tools and molecular breeding are here to stay if not indispensable in oil palm breeding. Oil palm breeding despite its long generation cycle disadvantage has been keeping abreast of the cross-cutting biotechnological advances from other annual and perennial crops. There will always be challenges and bottlenecks which could be solved or circumvented through further R&D. While undoubtedly biotechnological tools will keep on improving, the ultimate success of their application lies in developing a set of good field trials to provide reliable data which allows the continuous development of new biotechnological tools and the validation of their applications in breeding improvement and these are under the purview of the field plant breeder!

8. References

[1] Corley R H V and Tinker P B 2016. *The Oil Palm.* 5th Edition. Wiley Blackwell.
[2] Turnbull N, Durand-Gasselin T and Cochard B 2017 PalmElit/CIRAD'S breeding program and genetic progress *Oil Palm Genetics and Genomics* ed. Soh A C, Mayes S, and Roberts J (Atlanta: CRC Press) pp 165-174
[3] Soh A C, Mayes S and Roberts J 2017 *Oil Palm Breeding, Genetics and Genomics* (Atlanta: CRC Press)
[4] Jones L H 1974 *Oil Palm News* **17** 1
[5] Rabechault H and Martin J P 1976 *C.R.Acad.Sci.Paris Ser.D* **283** 1735
[6] Soh A C, Wong G, Tan C C, Chew P S, Chong S P, Ho Y W, Wong C K, Choo C N, Nor Azura H and Kumar K 2011 *J. Oil Palm Res.* **23** 935
[7] Choo C N, Wong W C, Wong C K, Tan C C and Soh A C 2017 Commercial clonal plant production. *Oil Palm Genetics and Genomics* ed. Soh A C, Mayes S, and Roberts J (Atlanta: CRC Press) pp 306-312
[8] Corley R H V, Lee C H, Law I H and Wong C Y 1986 *Planter* **62** 233
[9] Soh A C 1986 *Oléagineux* **41** 51
[10] Potier F, Nouy B, Flori A, Jacquarmard J C, Edyana Suryna H and Durand-Gasselin T 2006 Yield Potential of Oil Palm (*Elaeis guineensis* Jacq) clones: Preliminary Results Observed in the Aek Loba Genetic Block in Indonesia. Paper presented at the *ISOPB Seminar on Yield Potential in Oil Palm* in Phuket, Thailand December 2006.
[11] Chan P L, Ma L S, Low E T L, Elyana M S, Ooi, L C L, Cheah S C and Singh R 2010 *E-Journal Biotechnol.* **13** 14
[12] Ooi S E, Choo C N, Ishak Z and Ong-Abdullah M 2012 *Plant Cell, Tiss. Organ Cult.* **110** 201
[13] Ting N C, Jansen J, Nagappan J, Ishak Z., Chin C W, Tan S G, Cheah S C and Singh R 2013 *PloS One* **8**
[14] Ong-Abdullah M, Ordway J M, Jiang N, *et al* 2015 *Nature* **525** 533
[15] Soh A C and Mahamooth T 2017 Breeding for sustainability and climate change effects in oil palm. *Oil Palm Genetics and Genomics* ed. Soh A C, Mayes S and Roberts J (Atlanta: CRC Press) pp 355
[16] Murphy D J 2006 *In Vitro Cell. Dev. Biol.-Plant* **42** 89
[17] Yunus A M M and Parveez G K A 2008a *E-Journal Biotechnol.* **11** 3
[18] Yunus A M M Chai-Ling H and Parveez G K A 2008b *J. Oil Palm Res.* **2**
[19] Nurfahisza A R, Rafiqah M A, Masani M YA *et al* 2014 *J. Oil Palm Res.* **26** 96
[20] Izawati A M D, Parveez G K A and Masani M Y A 2012 *Methods in Molecular Biology* **847** 177
[21] Masani M Y A, Noll G A, Parveez G K A *et al* 2014 *PloS One* **9**
[22] Parveez G K A, Bahariah B, Ayub N H *et al* 2015 *Frontiers in Plant Sci.* **6** 598
[23] Izawati A M D, Masani M YA, Ismanizan I and Parveez G K A 2015 *Frontiers in Plant Sci.* **6** 727
[24] Murphy D J 2017 Transgenic oil palm *Oil Palm Genetics and Genomics* ed. Soh A C, Mayes S and Roberts J (Atlanta: CRC Press) pp 363-373.
[25] Mao Y, Zhang H, Xu N *et al* 2013 *Molecular Plant* **6** 2008
[26] Singh R, Low E T L, Ooi L C L *et al* 2013a *Nature* **500** 340
[27] Singh R, Ong-Abdullah M, Low E T L *et al* 2013b *Nature* **500** 335
[28] Singh R, Low E T L, Ooi L C L *et al* 2014 *Nature Comm.* **5** 1
[29] Wong W C, Teo C J, Wong C K, Mayes S, Singh R and Soh A C 2015 *J. Oil Palm Res.* **27** 113
[30] Wong C K and Soh A C 2017 Commercial Seed Production *Oil Palm Genetics and Genomics* ed. Soh A C, Mayes S and Roberts J (Atlanta: CRC Press) pp 296-306
[31] Mayes S, Jack P L, Corley R H V and Marshall D F 1997 *Genome* **40** 116

[32] Barcelos E, Amblard P, Berthaud J and Seguin M 2002 *Pesquisa Agropecuaria Brasileira* **37** 1105

[33] Rajanaidu N, Mohd Din, Marhalil M, Norziha A, Meilina O A, Fadila A M, Nor Azwani A B, Adelina L, Zulkfli A and Kushairi A 2017 Prospection, conservation and the broadening of the genetic base in Oil Palm *Oil Palm Genetics and Genomics* ed. Soh A C, Mayes S and Roberts J (Atlanta: CRC Press) pp 27-44

[34] Bernardo R 2014 *Crop Sci.* **54** 68

[35] Dunwell J M, Wilkinson M J, Nelson S, Wening S, Sitorus A C, Mienanti D, Alfiko Y, Croxford A E, Ford C S, Forster B P and Caligari P D S 2010 *BMC Plant Biol.* **10** 218

[36] Billotte N, Marseillac N, Risterucci A M, Adon B, Brotteir P, Baurens F C, Singh R, Herran Asmady H, Billot C, Amblard P, Durand-Gasselin T, Courtois B, Asmono D, Cheah S C, Rohde W, Ritter E and Charrier A 2005 *J. Oil Palm Res.* **17** 64

[38] Morcillo F, Cros D, Billotte N, Ngando-Ebongue, G F, Domonhédo H, Pizot M, Cuéllar T, Espéout, S, Dhouib R, Bourgis F, Claverol S, Tranbarger T J, Nouy B and Arondel V 2013 *Nature Comm.* **4** 2160

[39] Roberts J, Soh A C and Mayes S 2017 Harvestable yield *Oil Palm Genetics and Genomics* ed. Soh A C, Mayes S and Roberts J (Atlanta: CRC Press) pp 94-99.

[40] Gibson G 2010 *Nature Genetics* **42** 558

[41] Goddard M E and Hayes B J 2007 *J. Animal Breeding Gen.* **124** 323

[42] Resende M F R, Munoz P, Resende M D V, Garrick D J *et al* 2012 *Genetics* **190** 1503

[43] Wong C K and Bernardo R 2008 *Theor. Appl. Genetics* **116** 815

[44] Teh C K, Kwong Q B, Ong A L, Mohamed M, Apparow S, Chew F T, Appleton D S and Harikrishna Kulaveerasingam H 2017 Application of genomic tools in oil palm breeding *Oil Palm Genetics and Genomics* ed. Soh A C, Mayes S and Roberts J (Atlanta: CRC Press) pp 246-256

[45] Teh C K, Ong A L, Kwong Q B, Sukganah A, Chew F T, Mayes S, Mohaimi M, Appleton A and Harihrishna K 2015 *Sci. Rep.* **6**

[46] Alizadeh F, Abdullah S N A, Khodavandi A, Abdullah F, Yusuf U K and Chong P P 2011 *J. Plant Physiol.* **168** 1106

[47] Heslot N, Akdemir D, Sorrells M and Jannink J L 2014 *Theoret. Appl. Gen.* **127** 463

[48] Cros D 2017 Genomic Selection for oil palm *Oil Palm Genetics and Genomics* ed. Soh A C, Mayes S and Roberts J (Atlanta: CRC Press) pp263-269

IOP Conf. Series: Earth and Environmental Science **183** (2018) 012003 doi:10.1088/1755-1315/183/1/012003

Transformation of oil palm calli using CRISPR/Cas9 System: toward genome editing of oil palm

Asmini Budiani[1*], R A Putranto[1], I Riyadi[1], Sumaryono[1], H Minarsih[1] and R Faizah[2]

[1]Indonesian Research Institute for Biotechnology and Bioindustry, Jalan Taman Kencana No.1, Bogor 16128, Indonesia
[2]Indonesian Oil Palm Research Institute, Jalan Brigjen Katamso No.51, Medan 20158, Indonesia

*E-mail: asminib@yahoo.com

Abstract. Genome editing using CRISPR/Cas9 construct containing sgRNA is the newest technology used to assemble organisms with desired characters through knock out of the target gene. As part of the effort to develop oil palm planting material tolerant to *Ganoderma*, this research aimed to obtain a technique for introducing CRISPR/Cas9 modules containing sgRNA into oil palm tissue. Based on differential expression studies, two genes encoding isoflavone reductase (IFR) and methallothionine-like protein (MT) were selected as the target for edited genes. The CRISPR-IFR and CRISPR-MT modules each carrying a sequence of gene recognition (sgRNA) were constructed and transformed into oil palm calli using *Agrobacterium tumefaciens*. The transformed calli were cultured on a modified DF medium for callus propagation, and then sub-cultured into DF media for induction of somatic embryos. PCR analyses using specific primers flanking the DNA targeted region were carried out and the results showed that both constructs had been introduced into the cells. DNA sequence analysis indicated the presence of substitution of several bases in the target area of both genes. The transformed calli grew in a selection medium containing hygromycin. Development of somatic embryos was detected after 3-4 weeks culturing on modified DF media.

1. Introduction

Basal Stem Rot (BSR) disease caused by *Ganoderma* sp. has greatly threatened the sustainability of oil palm plantations especially in Indonesia and Malaysia [1]. The infected plants are usually dead; consequently, decrease the number of plants per unit area which in turn decrease the productivity and cause huge losses. At an average attack rate of 1%, the value of losses could reach 256 million USD per year [2]. Many efforts have been conducted to overcome this disease; unfortunately, the decline of Ganoderma attack has not been seen. On the contrary, the number of affected plants is reported to be increasing year to year. The use of planting material tolerant to Ganoderma is believed to be the most potential approaches to overcome this problem.

Cisgenesis, one of the modern biotechnology-based approaches, is claimed to be a potential technology for creating new varieties of plant in relatively quick and precise manner. Since the cassette of DNA used for modifying the targeted genome is derived from the same plant, the modified targeted plant is not categorized as a transgenic plant [3]. Genome editing, one of cisgenic technology, is the most prospective technology to replace random mutagenesis methods, such as ethyl methanesulfonate (EMS) mutagenesis. Genome editing allows proper genetic manipulation of specific genome sequences [4, 5]. The ideal scenario of genome editing is to minimize insertion of the non-crop genetic sequences into host plant genomes by cutting out the entire non-crop DNA sequences such as gene markers for selection, Cas proteins, etc., after the target gene edited [6].

There are three modules of DNA cleavage, (1) Zinc Finger Nucleases (ZFNs), (2) Transcription Activator-Like Effector Nucleases (TALENs) and (3) The immune system adaptive prokaryotic type II CRISPR (Clustered Regularly Interspaced Short Palindromic Repeats)/Cas (CRISPR-associated) [4, 7, 8]. Both nucleases (ZFNs and TALENs) are used to mutate the plant genome at the specific locus. However, these systems have limited applications because they require two different DNA protein binding and flanking sequences, each with a module nuclease FoKl [4, 9]. Compared to ZFNs and TALENs, the CRISPR/Cas9 module is simpler and considered as a breakthrough technique for genome editing. The module is based on Cas9 nuclease and a modified RNA single guide (sgRNA) to edit target nucleotide sequences [10]. These systems rely on sequence-specific nucleases (SSNs) to produce double-strand breaks (DSBs) at specific locations within the genome. DBS is repaired by the cell using a mechanism of non-homologous end joining (NHEJ) or homologous repair (HR) [9, 11].

Applications of CRISPR/Cas9 system have been reported in several plant species such as *Arabidopsis thaliana*, *Nicotiana benthamiana*, *Oryza sativa*, *Triticum aestivum*, *Solanum lycopersicum*, and *Sorghum bicolour* [4, 9]. By the end of 2014, genome editing has also been successfully applied on citrus and tomato plants [12, 13]. So far, there was no report on the application of CRISPR/Cas9 on genome editing of oil palm. As a part of the effort to develop oil palm planting material tolerant to Ganoderma through genome editing, this research aims to establish technique for delivering CRISPR/Cas9 module containing sgRNA for the model genes, namely metallothionein-like protein (MT) and isoflavone reductase (IFR) into oil palm calli and to regenerate the transformed calli into somatic embryos.

2. Materials and Methods

2.1. Callus induction from oil palm young leaf explant

Young leaf pieces were cultured on MS modified solid medium containing auxin and cytokinin for callus induction according to the method described by Sumaryono *et al.* [14]. The culture was then incubated in a dark room at 26° C. Initial callus began to appear after 4 to 6 months. Callus proliferation was then conducted by sub culturing every 8 weeks on the same medium with a lower auxin concentration. This callus was then used for transformation using the CRISPR/Cas9 module.

2.2. Construction of CRISPR/Cas9 system with targeted model genes on oil palm

Two genes were selected as a target for genome editing, namely isoflavone reductase (IFR) and methallotioneine like protein (MT). Construction of CRISPR/Cas9 system specifically targeted for each gene was initiated by determining the target site (also known as protospacer). From the site, 20 nucleotides were selected following the criteria of Hwang *et al.* [15] for designing primer of the target sequence. The design of single guide RNA (sgRNA) and Cas9 proteins was conducted using Geneious software following modified protocol of Ran *et al.* [16]. If the target DNA had more than one isoform with sequence differences, chimeric guide RNAs (sgRNAs) would be designed to target 1-2 gene model sites.

Table 1. Primers used in this experiment with its DNA sequences.

Primer Name	DNA Sequences
CRISPR-F	5' - GACCAAGCCCGTTATTCTGAC-3'
CRISPR-R	5'- AAGTCTGATGCAGCAAGCGAG-3'
NPTII-F	5'- GAACAAGATGGATTGCACGC-3'
NPTII-R	5'- GAAGAACTCGTCAAGAAGGC-3'
HYG-F	5'-ACTATCGGCGAGTACTTCTACA-3'
HYG-R	5'-GTATCACTGGCAAACTGTGATG-3'
sgRNA1 (IFR)	5'-GTGCAACGTGTACCCTGATCtgg-'3'
sgRNA2 (MT)	5'-GAAGCTCCTGGACGACTTCAagg-3'
Isoflavone reductase II (*EgIFR-F2*)	5'-ACCGGGTACATCGGAAAGTT-3'
Isoflavone reductase II (*EgIFR-R2*)	5'-GGAGATCACCGTGTCCACTT-3'
Isoflavone reductase I (*EgIFR-F1*)	5'-GATCATCGCCGCAATAAAAG-3'
Isoflavone reductase I (*EgIFR-R1*)	5'-ATGGGAGGAAGTAACCAGCA-3'
Metallothionein-like protein I (*EgMT-F1*)	5'-AGGCAAATGTGGCTGTGGCGTT-3'
Metallothionein-like protein I (*EgMT-R1*)	5'-ACTTGCAGTTGCAGCCTCCGTT-3'
Metallothionein-like protein II (*EgMT-F2*)	5'-GGGCACTATGAAGGGTTTGA-3'
Metallothionein-like protein II (*EgMT-R2*)	5'-TTGGATGCTTGGAAGGAGAC-3'

2.3. Establishment of transformation methods using Agrobacterium tumefaciens *carrying CRISPR/ Cas9 module*

DNA transformation of oil palm callus was carried out using *A. tumefaciens* system carrying the CRISPR/Cas9 constructs, each with sgRNA specifically targeted for MT or IFR genes (CRISPR-MT and CRISPR-IFR). Each construct was first introduced into *A. tumefaciens*. In addition, both constructs were also cloned into *E. coli*. The transformation process was done by immersing the oil palm callus along with *A. tumefaciens* for 15 min. The callus was then cultured on the co-cultivation media containing100 ppm acetosiringone for two days followed by culture on media containing 250 ppm cefotaxime for 5 days. The calli were then transferred into selection media containing and 25 ppm hygromycin and 250 ppm cefotaxime for 3-4 weeks. The transformed calli were cultured on the media for callus multiplication. Successful introduction of CRISPR-MT and CRISPR-IFR into *A. tumefaciens* and *E. coli* was confirmed by PCR using NPTII primers, HYG primers and CRISPR monocot-specific primers, followed by DNA analysis. The callus was then sub cultured into the media for regeneration of somatic embryos. DNA sequences of the primers used in this research were shown in Table 1.

3. Results and Discussion

3.1. Introduction of CRISPR constructs into A. tumefaciens *and* E. coli

Figures 1 and **2** illustrate the electrophoresis of colony PCR products of *E. coli* and *A. tumefaciens* transformed using CRISPR-MT and CRISPR-IFR constructs. These results showed that both constructs have been introduced into *A. tumefaciens* and cloned into *E. coli*. The *A. tumefaciens* containing CRISPR constructs with sgRNA of MT and IFR genes were then used in the transformation of oil palm calli for editing MT and IFR genes.

IOP Conf. Series: Earth and Environmental Science **183** (2018) 012003 doi:10.1088/1755-1315/183/1/012003

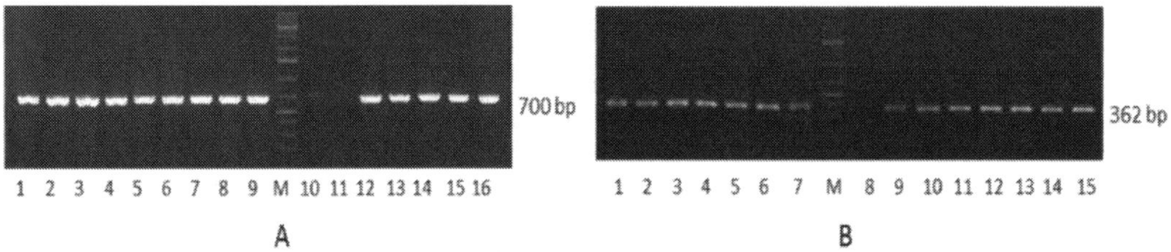

Figure 1. (A) Colony PCR using NPTII primers of the transformed *E. coli*. Lanes (1-9) colonies transformed using CRISPR-MLP: (10) Control pCAMBIA 1303 P5CS, (11) negative control, (12-16) colonies transformed with CRISPR-IFR. (B) Colony PCR using CRISPR primers: lanes (1-6) colonies transformed with CRISPR-MT, (7) plasmid DNA CRISPR-MT, (8) negative control, (9) plasmid DNA CRISPR-IFR, (10-15) colonies transformed with CRISPR-IFR, (M) Marker 1 kb plus ladder.

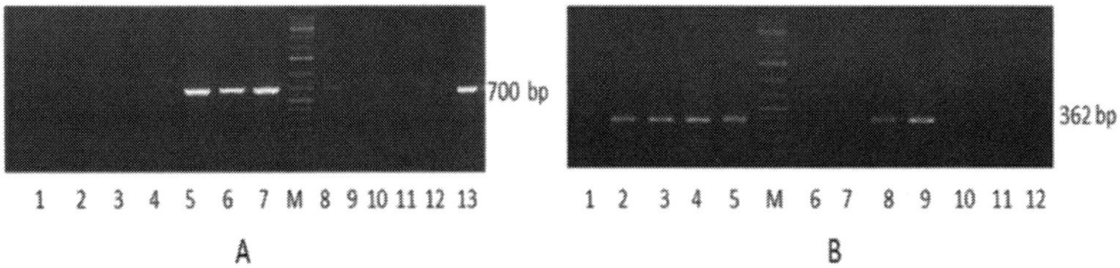

Figure 2. (A) Colony PCR using NPTII primer of *A. tumefaciens* transformed using CRISPR modules. Lanes (1-7) colonies transformed by CRISPR-IFR construct, (8) Control pCAMBIA 1303 P5CS, (9) Negative control, (10-13) colonies transformed by CRISPR-MT construct, (B) Colony PCR of *A. tumefacien* using CRISPR primer. (1-4): colonies transformed with CRISPR-IFR construct, (5) plasmid DNA of CRISPR-IFR, 6) negative control, (7) control of Plasmid DNA LBA4404, (8) plasmid DNA of CRISPR-MT, (9-12) colonies transformed with CRISPR-MT construct, (M) 1 kb plus DNA ladder.

3.2. Transformation of oil palm calli using CRISPR-IFL and CRISPR-MT constructs

Development of oil palm calli transformed with both CRISPR/Cas9 constructs into somatic embryos is presented in **Figure 3**. It is shown that on the media containing hygromycin as selected marker, callus transformed with CRISPR construct developed into somatic embryos (Figure 3E and 3F). On contrary, the untransformed calli did not develop and tend to browning. This result, followed by PCR analysis using specific genes (IFR and MT) and hygromycin primers prove that both constructs have been introduced into oil palm genome (**Figure 4**).

To identify the occurrence of mutation on the sgRNA targeted DNA region, the DNA sequences of these PCR products were aligned with that of the control untransformed DNA. **Figure 5** show that introduction of CRISPR-IFR2 construct (A1-A6) induce InDels mutation causing changes of amino acid sequences which may alter the function of its proteins. It can be seen on A3, A4 and A5 lines that there is nucleotide substitution, resulted in CAGCTT---GG-AA-T sequence compared with the untransformed control, AGCTCC---AC-CT-C-A. In those three lines, deletion was also occurred in the 111 to 115 nucleotide sequences. On the other side, there is no mutation detected on the A1, A2, and A6 lines. When viewed from the protein translation, substitutions on A3, 4 and 5 change the sequence of amino acid residues, i.e. TALGEIRVRV compared with the untransformed control, KLLDDFKASGV. Changes the sequence of amino acid residues is confirmed to have an effect on the function of the IFR2 protein, however further proof has to be observed on the developing transformed callus.

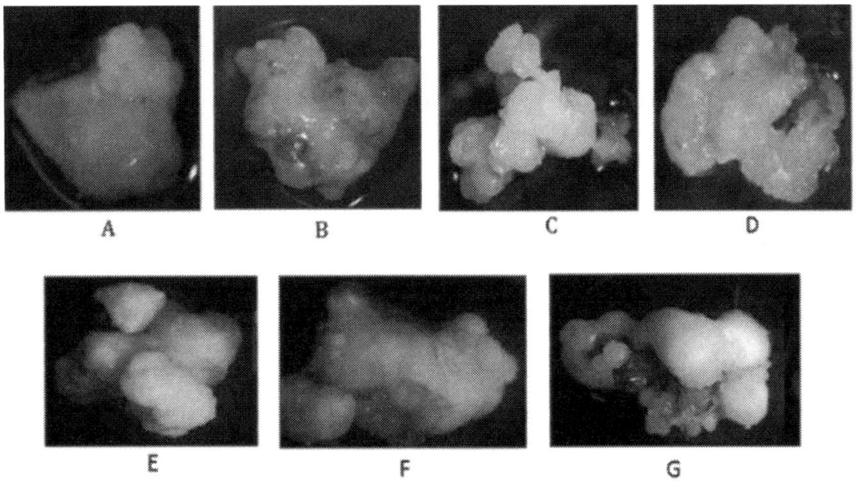

Figure 3. Oil palm calli 5 days after transformation using CRISPR-IFR (A) and using CRISPR-MT (B), untransformed callus on the media without antibiotic (control) (C), untransformed callus on the media with antibiotic (negative control) (D), somatic embryo developed from the CRISPR-IFR–transformed callus (E), somatic embryo developed from the CRISPR-MT–transformed callus (F), and somatic embryo developed from untransformed callus on the media without antibiotic (positive control) (G).

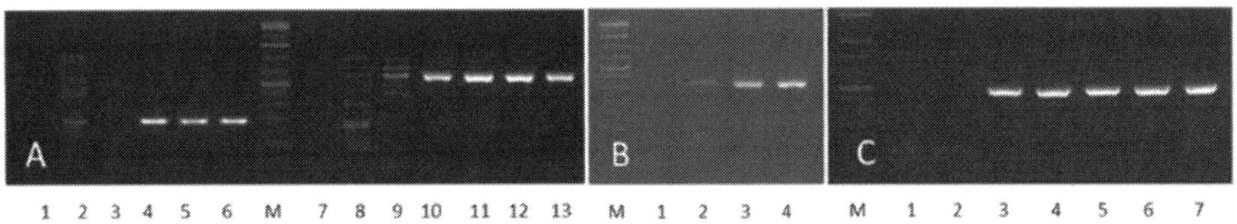

Figure 4. PCR of oil palm calli using primer of MT2 (1-6) dan MT1 (7-13) (A). Lanes 1,7: negative control, 2, 8, & 10: untransformed calli, 3 & 9 plasmid DNA of CRISPR-MT, 4-6, & 11-13 calli transformed by CRISPR-MT; using IFR2 primers (B). Lane 1: Negative control, 2: untransformed callus, 3&4 calli transformed by CRISPR-IFR; hygromycin primers (C). Lane 1: untransformed callus, 2: negative control, 3&4 calli transformed by CRISPR-IFR, 5-7: calli transformed by CRISPR-MT; (M) 1 kb plus DNA ladder.

DNA alignment of the sgRNA target region of the transformed oil palm calli using CRISPR-MT constructs showed that introduction of CRISPR-MT construct induced mutation in the sgRNA region which may cause the MT protein to no longer function. DNA alignment is carried out between untransformed callus (NT-MT) and transformed callus (B1, 7, 9, and 10) which may be considered as a new strain. It can be seen in **Figure 6** that the most basic substitution occurs in the B9 strain with the change of GAT-A-ATA-T------T---CC which converts the translation of the protein into IKINVYSALG compared with the untransformed control, RCNVYPDLGG. Individual substitution occurs in the B7 strain that changes the residual amino acid N (on the untransformed calli/control) to G. The basic substitution in line B10 converts the amino acid residue R to W and V to A. In the B1 strain, the substitution occurs at the beginning of sgRNA, nucleotide bases AG (on the untransformed calli/controls) to TA causes the stop codon to occur right at the end of sgRNA. From this result it can be ascertained that the B1 strain received the most significant impact of the knock-out system. The MT protein was definitely not working properly on the B1 strain.

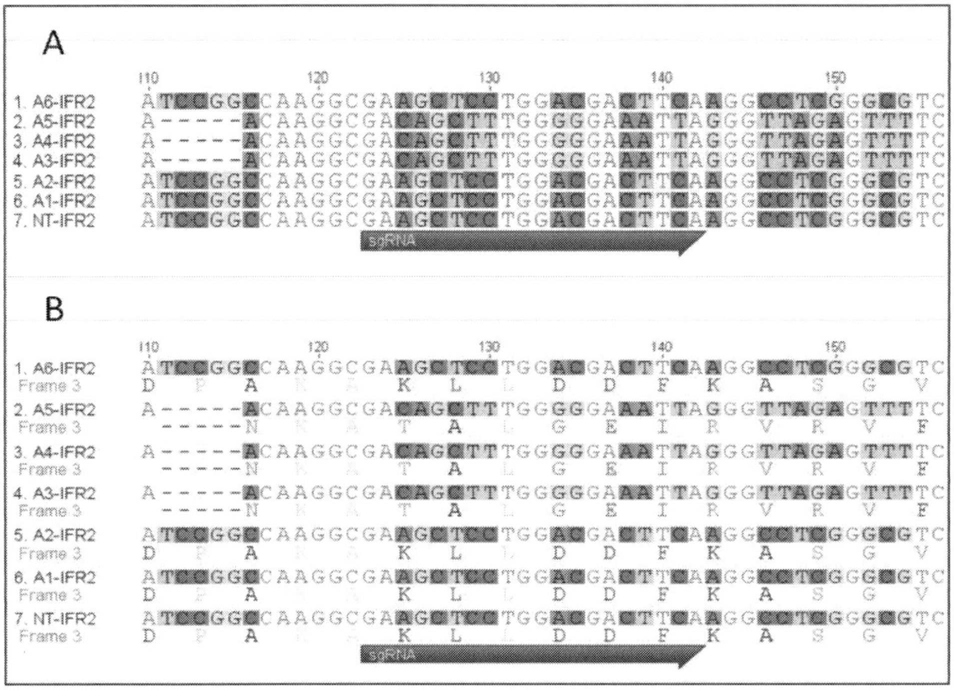

Figure 5. Alignment of sgRNA area from oil palm calli transformed by CRISPR-IFR2 module compared with the untransformed callus. (A) DNA sequences. (B) DNA and amino acid sequences. NT: untransformed callus. A1-6: transformed calli from 6 different lines.

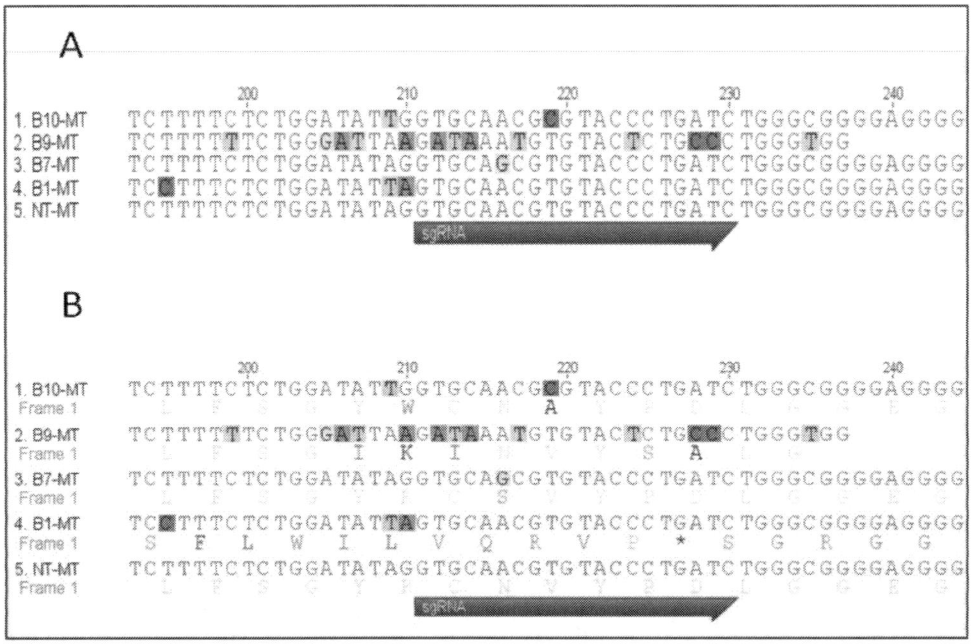

Figure 6. Alignment of sgRNA area from oil palm calli transformed by CRISPR-IFR2 module compared with the untransformed callus. (A) DNA sequences. (B) DNA and amino acid sequences. NT: untransformed callus. A1-6: transformed calli from 6 different lines.

4. Conclusions

CRISPR/Cas9 constructs carrying sgRNA with targeted MT and IFR genes were successfully introduced into oil palm callus caused mutations in both targeted DNA regions resulting in alterations on sequences of the amino acid residues. On the selection medium, the transformed calli developed into somatic embryos.

5. Acknowledgments

This work was financially supported by the Indonesian Estate Crop Fund for Oil Palm (BPDPKS). Authors thanks to Dr. Priyono, DIRS, Director of Indonesia Research Institute for Biotechnology and Bioindustry, and to Dr. Gede Wibawa, R&D Director of PT RPN for supporting this research.

6. References

[1] Hushiarian R, Yusof N A and Dutse SW 2013 *SpringerPlus* **2** 555
[2] Taniwiryono D 2010 *Sinar Tani* **11** 20
[3] Kanchiswamy C N, Malnoy M, Velasco R, Kim J-S and Viola R 2015 *Trends in Biotechnol.* **33** 489
[4] Belhaj K, Chaparro-Garcia A, Kamoun S and Nekrasov V 2013 *Plant Methods* **9** 39
[5] Belhaj K, Chaparro-Garcia A, Kamoun S, Patron N J and Nekrasov V 2015 *Curr. Opinion in Biotechnol.* **32** 76
[6] Sticklen M 2015 *Adv. Crop Sci. Tech.* **3** 123
[7] Puchta H and Fauser F 2013 *The Inter. J. Dev. Biol.* **57** 629
[8] Sprink T, Metje J and Hartung F 2015 *Curr. Opinion in Biotechnol.* **32** 47
[9] Bortesi L and Fischer R 2015 *Biotechnol. Adv.* **33** 41
[10] Briner A E, Donohoue P D, Gomaa AA, Selle K, Slorach E M, Nye C H, Haurwitz R E, Beisel C L, May A P and Barrangou R 2014 *Mol. Cell* **56** 333
[11] Voytas D F and Gao C 2014 *PLoS Biol* **12** e1001877
[12] Jia H and Wang N 2014 *PLoS ONE* **9** e93806
[13] Martinelli F, Grillone G and Sgroi F 2014 *American J. Applied Sci.* **11** 1904
[14] Sumaryono, Riyadi I, Kasi P D and Ginting G 2008 *Indon. J. Agric.* **1** 109
[15] Hwang W Y, Fu Y, Reyon D, Maeder M L, Tsai S Q, Sander J D, Peterson R T, Yeh J R J and Joung J K 2013 *Nat. Biotechnol.* **31** 227
[16] Ran F A, Hsu P D, Wright J, Agarwala V, Scott D A and Zhang F 2013 *Nat. Protocols* **8** 2281

IOP Conf. Series: Earth and Environmental Science **183** (2018) 012004 doi:10.1088/1755-1315/183/1/012004

Semi-quantitative RT-PCR analysis of transcripts encoding protease inhibitor in *Hevea brasiliensis* Muell. Arg latex

Irfan Martiansyah[1], D M Amanah[1] and R A Putranto[1*]

[1]Indonesian Research Institute for Biotechnology and Bioindustry, Jalan Taman Kencana No.1, Bogor 16128, Indonesia

*E-mail: rizaputranto@iribb.org

Abstract. In this paper, the semi-quantitative RT-PCR was carried out on four *Hevea brasiliensis* protease inhibitor transcripts (*HbPI1-4*) of rubber tree clone PB 260 under abiotic and biotic stresses. The research aimed to understand whether the expressions of HbPIs are induced by either biotic or abiotic stresses. The RNA isolation was performed under strict procedure to synthesize the best quality of cDNA. The accumulation of gene transcript was determined by RT-PCR (Reverse Transcriptase – Polymerase Chain Reaction) at the 30th cycle. Pre-PCR experiments were carried out on 20th, 25th, 30th, and 35th cycles for each gene. At the 30th cycle, PCR products were clearly visible on an agarose gel and could be quantified in comparison to the reference gene (*HbActin*). In general, the value of gene expression of *HbPIs* ranged from 0.222 to 0.531 in various conditions. However, the data were not statistically different under abiotic (tapping panel dryness, TPD) and biotic (white root disease, WRD) stresses. The semi-quantitative RT-PCR was able to confirm the relatively stable expression of *HbPIs* under five tree conditions (healthy trees, TPD-affected trees at <50% dry cut length (DCL), TPD-affected trees at >50% DCL, WRD-affected trees and both of symptoms).

1. Introduction

The milky white latex is the main product of the para rubber tree, *Hevea brasiliensis* Muell. Arg [1]. It contains the natural rubber commercially used as the raw material of industrial products such as tire and rubber bearing due to its unique physical properties such as elasticity, impact resistance and abrasion [2, 3]. Latex is the cytoplasm of highly differentiated cells namely laticifers [4, 5]. Latex is harvested by tapping the soft bark truncating the laticifers. The latex flows out until the beginning of the coagulation processes plugs the severed vessels [6, 7].

Inside latex, various types of protein have been detected in which they have associated functions in plant defense against pathogens and wounding responses [8-10]. In a matter of fact, twenty-five per cent of the proteins in C- and B-serums are known as antimicrobial agents in *Hevea brasiliensis*. Most of these proteins have the activity of chitinase/lysozyme which can degrade fungal and bacterial cell wall components such as chitin and peptidoglycan, respectively [8, 11]. However, among these known proteins, small size proteins known as protease inhibitor were less studied in *Hevea* latex.

Limited numbers of research have reported the isolation and characterization of protease inhibitor from the latex of two *Hevea* tree clones RRIM 600 and PB 260 [11-15]. In molecular levels, high numbers of Expressed Sequence Tags (ESTs) encoding protease inhibitors were predicted in laticifers

[10]. Taken together, the studies showed that the protease inhibitors were encoded by numerous genes. A comparative genome analysis between *Arabidopsis thaliana* and *Hevea brasiliensis* has been carried out to putatively identify the gene family encoding *Hevea* protease inhibitors (HbPIs) [16]. Further transcript quantification analyses are necessary to confirm the presence of these genes in the latex. In addition, the expression analysis should be carried out during contrasted condition such as diseased and healthy rubber trees to determine the constitutive or inducible expression of HbPI encoding genes.

Nowadays, the accumulation of messenger RNA (mRNA) in the plant can be measured using a rapid, sensitive and reliable method [17-19]. Despite the greater accuracy of recently developed techniques such as real-time PCR, semi-quantitative PCR methods are still widely used and appropriate for many purposes in plant studies [20-23]. The main reason for its popular use is that the semi-quantitative method provides a combination of traditional PCR and transcript quantification at a relatively low cost [22]. The semi-quantitative PCR was equally confirmed to generate results as well as the more advanced methods [24].

In this paper, an alternative semi-quantitative RT-PCR method for comparing the expression of four transcripts encoding putative *Hevea brasiliensis* protease inhibitors under the abiotic and biotic stresses was presented. The research aimed to understand whether the expressions of HbPIs are induced by WRD, TPD or combination of both. Firstly, the RNA isolation was performed under strict procedure to synthesize the best quality of cDNA. Secondly, the accumulation of gene transcript was determined by RT-PCR. Thirdly, the quantification of transcript expression was normalized using the expression of a reference gene on a gel electrophoresis analyzer. Fourthly, data and statistical analysis were carried out.

2. Materials and Methods

2.1. Plant material

Latex samples were obtained from rubber tree clones PB 260 growing at the field trial of Sembawa Rubber Research Institute, Palembang, Indonesia (Latitude: S 2° 55' 38.687"; Longitude: E 104° 32' 18.918"). The sampling of latex was carried out at 7 to 8 am to obtain a sufficient amount of latex. The samples were collected in five conditions i.e. healthy trees, tapping panel dryness (TPD)-affected trees at <50% dry cut length (DCL), TPD-affected trees at >50% DCL, white root disease (WRD)-affected trees and both of symptoms (**Figure 1**). TPD as a physiological disease represents abiotic stress while WRD represents biotic stress. The determination of the tapping panel dryness syndrome followed the measurement of DCL [25]. The collected samples were brought to the laboratory on dry ice and stored in a deep freezer (-80°C) until RNA extraction.

2.2. RNA extraction and preparation

The RNA was extracted using the modified method of Chang [26]. The latex samples were briefly thawed in a water bath at 50°C and then transferred to centrifuge tubes. The thawed latex was centrifuged for 40 min at 15,000 g at 20°C. The white fraction was collected and transferred to new centrifuge tubes. Series of organic solution addition and centrifugation was performed. First, one volume of phenol/chloroform/isoamylalcohol (25/24/1; v/v/v) was added and mixed at room temperature followed by a centrifugation for 15 min at 10,000 g at 4°C. The resulted upper phase (aqueous) was collected into new centrifuge tubes. Secondly, one volume of phenol/chloroform/ isoamylalcohol (25/24/1; v/v/v) was added and mixed at room temperature followed by a centrifugation for 15 min at 10,000 g at 4°C. The upper phase (aqueous) was collected into new centrifuge tubes. Thirdly, one volume of chloroform/isoamylalcohol (24/1; v/v) was added and mixed at room temperature followed by a centrifugation for 15 min at 10,000 g at 4°C. An addition of 8M LiCl to the resulted aqueous phase was carried out to reach a final concentration of 2M. The solution was mixed by inverting tubes several times and incubated overnight at 4°C. The mixture was then centrifuged for 30 min at 10,000 g at 4°C. The supernatant was removed. A 400 μL of DEPC water was added and agitated periodically to help dissolve the RNA pellet. The RNA solution was transferred into 1.5 mL microtubes. A volume of 400 μL phenol/chloroform/isoamylalcohol (25/24/1;

v/v/v) was added and mixed followed by a centrifugation for 15 min at 13,000 g at 4°C. One volume of chloroform/isoamylalcohol (24/1; v/v) was added into the upper phase followed by a centrifugation for 10 min at 10,000 g at 4°C. The aqueous phase was collected. A 1/10 volume of sodium acetate 3M pH 5.2 and 3 volume of absolute ethanol were added to precipitate the RNA for 3 hours in a freezer. After centrifugation for 30 min at 13,000 g at 4°C, the supernatant was discarded. The pellet was rinsed with 70% ethanol and centrifuged for 5 min at 13,000 g at 4°C. The pellet was dried in a speed vacuum and resuspended in 30 μL of nuclease-free water.

To eliminate potential DNA contamination, total RNA was treated with DNase I (Sigma-Aldrich, USA). A volume of 1 μL DNase I was added to 8 μL RNA (1000 ng) for each sample. After incubation for 15 min, the RNA was dissolved in 1 μL STOP solution (50 mM EDTA) for 10 min at 70°C. Agarose gel electrophoresis was used to visualize the RNA integrity while the quantity of RNAs was measured using NanoDrop 2000 (ThermoScientific, USA). RNAs were conserved at -80°C.

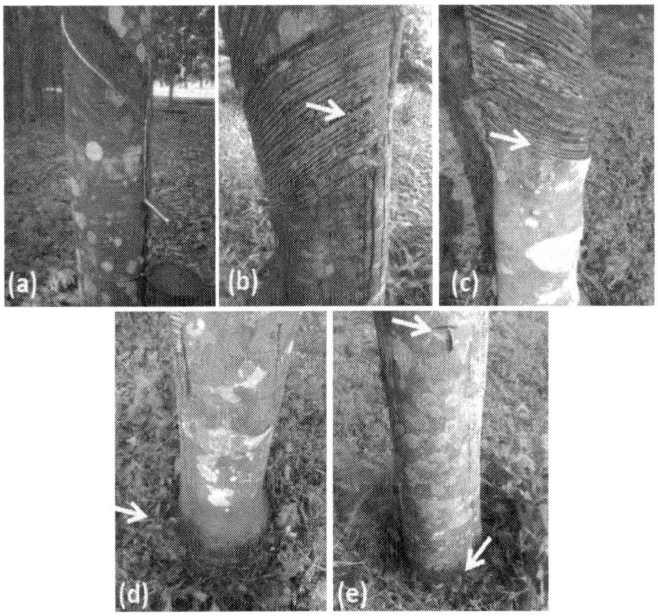

Figure 1. Illustration of rubber tree clones PB 260 in the sampled field trial. (a) healthy bark (normal latex flow); (b) TPD-affected bark at ≤50% DCL (arrow: partial cessation of latex flow); (c) TPD-affected bark at >50% DCL (arrow: complete cessation of latex flow); (d) WRD-affected tree (arrow: fruit body of *Rigidosporus lignosus*); (e) WRD+TPD (arrow: complete bark dry and fruit body of the infectious fungi). WRD: white root disease; TPD: tapping panel dryness; DCL: dry cut length.

2.3. cDNA synthesis and semi-quantitative RT-PCR procedures

The semi-quantitative RT-PCR analysis was conducted as follows. cDNAs were synthesized from 1 μg of total RNA to the final 20 μL reaction mixture using Bioneer AccuPower Cycle Script RT PreMix kit (Bioneer, USA) according to the manufacturer's instructions. Full-length cDNA synthesis was checked on each cDNA sample by PCR amplification of the Actin cDNA using primers at the cDNA ends. Reverse Transcriptase (RT)-PCR was finally carried out using a thermocycler. The standard program comprised a cycle of 10 min at 95°C, 45 sec at 65°C and 60 sec at 72°C, followed by 45 sec at 95°C, 45 sec at 65°C and 60 sec at 72°C for 30 cycles. The 30th cycle was selected after pre-PCR on several cycles (20th, 25th, 30th, and 35th) for each gene. At the 30th cycle, the PCR product was clearly visible on an agarose gel and could be quantified in comparison to the reference gene

(*HbActin*). The primer pairs used in this study were as follow: *HbActin* (Genbank accession: JF270598.1) with 5'-AGTGTGATGTGGATATCAGG-3' (forward) and 5'-GGGATGCAAGG ATAGATC-3' (reverse), *HbPI1* (Genbank accession: KX101219.1 with 5'-GGGGAATCCACA GATTTCATT-3' (forward) and 5'-TTTAGCCAATGGCAGGAACT-3' (reverse), *HbPI2* (Genbank accession: KX101220.1) with 5'-GAATGGCAAGTCAGTGTCCA-3' (forward) and 5'-TTTAGCCAATGGCAGGAACT-3' (reverse), *HbPI3* (Genbank accession: KX101221.1) with 5'-TCGAGAGGAGAATGGCAAGT-3' (forward) and 5'-GAAAACCAGGACCCTGCATA-3' (reverse), *HbPI4* (Genbank accession: KX101222.1) with 5'-AGTGTGATGTGGATATCAGG-3' (forward) and 5'- GGGATGCAAGGATAGATC-3' (reverse).

2.4. Acquisition of gel images and quantitative analysis

Fifteen μL of the PCR products were resolved by electrophoresis on ethidium bromide-stained gels, scanned using GelDock Imaging System (California, USA). The band quantification was carried out by scanning densitometry using Gel Analyzer 2010 (University of Leicester, UK). The measurement resulted in band intensity as the expression unit. The ratio between the sample RNA to be determined and *HbActin* was calculated to normalize for initial variations in sample concentration. Mean and standard deviation were calculated after normalization to *HbActin*. Relative expression (RE) of target genes against internal control was calculated by the equation: RE = expression unit of target genes/expression unit of internal control.

The mean values of normalized expression were obtained from three biological replicates. Statistical analysis was performed using the Analysis of Variance (ANOVA) at 5% level followed by Newman-Keuls Student test (XLStat, Addinsoft, USA).

3. Results and Discussion

3.1. Determination of RNA quality and PCR cycle for observation

The validation of RNA integrity is essential prior to the analytical pipeline for transcript accumulation analysis [27]. The quality and quantity of RNA determine largely the quality and quantity of synthesized cDNAs [28]. In this work, total RNA was isolated from every 5 samples of healthy and disease-affected rubber trees (**Figure 2a**).

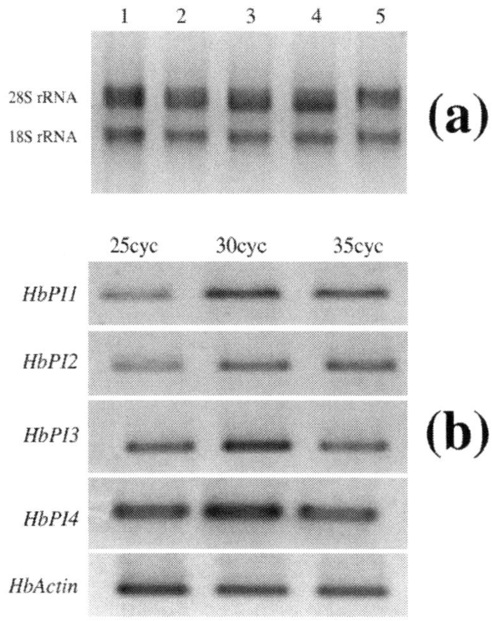

Figure 2. RNA visualization (a) and determination of the exponential range of amplification at 25th, 30th and 35th cycles for four transcripts encoding *Hevea* protease inhibitors. (b) (1) healthy bark; (2) TPD-affected bark at ≤50% DCL; (3) TPD-affected bark at >50% DCL; (4) WRD-affected bark; (5) WRD+TPD-affected bark.

The agarose gel visualization showed a good RNA integration with the presence of 18S and 28S rRNAs. The purity of the total RNA samples (ratio of 260/280 nm) ranged from 1.79 to 1.95 while the RNA concentrations ranged from 600 to 990 ng/μL. According to the literature, the use of partially damaged or DNA contaminated RNA, approximately 7% of the transcript expression led to a less optimum result [29]. This could mean that the result of fold change expression analysis comparing two opposed conditions could result in a false positive [27]. In this work, the total RNA was ensured to be free from DNA contamination by carrying out a procedure of DNase.

One of the most important rules of thumb when performing a semi-quantitative RT-PCR is to select the appropriate number of cycles. This determination of the optimum range of amplification is indispensable (1) to have a clearly visible band on agarose gel and (2) to be able to be quantified using a visualization software. The optimal number of cycles on both internal control and the target RNA has to be in the same range so that both can be measured on the same gel [22]. In this work, the determination of optimum PCR cycle was carried out on 20th, 25th, 30th, and 35th cycles (**Figure 2b**). On the 20th cycle, the gel was unclear and cannot be visualized. However, the band was visually clear on the 30th cycle for all transcripts *HbPIs* and equally *HbActin* so that this cycle was selected for semi-quantitative RT-PCR visualization.

3.2. Semi-quantitative RT-PCR

In this paper, the semi-quantitative RT-PCR was successfully carried out on five cDNA samples representing healthy and disease-affected rubber tree latex. With each of the chosen RNA sample, 258-bp fragment of *HbPI1*, 216-bp of *HbPI2*, 179-bp of *HbPI3*, 312-bp of *HbPI3*, and 188-bp of *HbActin* were specifically amplified. Using the *HbActin* as an internal control, the mRNA levels of the above four transcripts were determined among various cDNA samples (**Figure 3**).

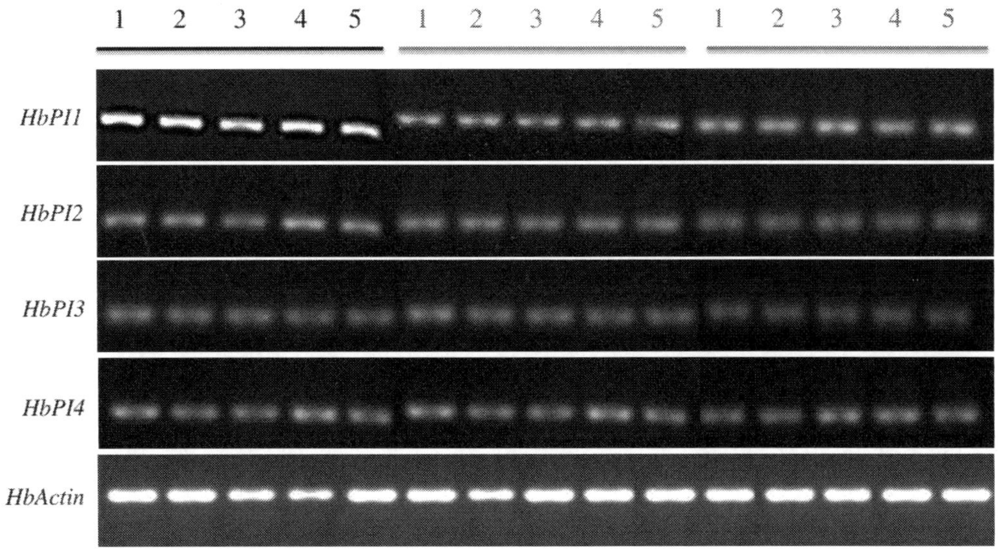

Figure 3. Semi-quantitative RT-PCR of four *Hevea brasiliensis* transcripts encoding protease inhibitors (*HbPIs*) under abiotic and biotic stresses. *HbActin* served as an internal control to normalize the expression levels of each *HbPI* transcripts. Each PCR was optimized for 30 cycles to achieve non-saturated gel image. Each colour bar represents one replicate. (1) healthy bark; (2) TPD-affected bark at ≤50% DCL; (3) TPD-affected bark at >50% DCL; (4) WRD-affected bark; (5) WRD+TPD-affected bark. WRD: white root disease; TPD: tapping panel dryness; DCL: dry cut length.

As shown in **Figure 3**, the band intensity of RNA transcripts of all test genes was visually lower than the internal control (*HbActin*). The expression of *HbActin* was stable throughout the different conditions of the cDNA sample which confirms the right choice of *HbActin* as the internal control. Meanwhile, the expression of all *HbPI* transcripts visually varied throughout the different sample conditions. The variation could lead to a potential assumption that *HbPI* transcripts were differentially regulated by abiotic and biotic stresses. After converting the visual image of agarose gel into semi-quantified data using Gel Analyzer, **Figure 4** confirmed the differential expression of *HbPI* transcripts among different cDNA samples. In general, the transcript accumulation of *HbPI1* and *HbPI2* were higher than *HbPI3* and *HbPI4* regardless of tree conditions (healthy or diseased). The combination of WRD and TPD had the highest relative expression of *HbPI2* transcript with the value of 0.531. On the other hand, *HbPI3* had the lowest relative expression on WRD-affected bark with the value of 0.222. However, the data was not statistically different under abiotic (TPD) and biotic (WRD) stresses. The mean values from three biological replicates showed a high variation. It means equally that the expression of *HbPI* transcripts was not regulated by either two diseases in a given time and tissue.

In the previous studies, the semi-quantitative RT-PCR analysis has been carried out in *Hevea brasiliensis*. The expression of *HbACO* genes in response to ethephon or ethylene was clearly observed using this technique [30]. The semi-quantitative RT-PCR has been successfully revealed the expression of the transcripts of two rubber biosynthesis genes (*HMGS2* and *HRT1*) and one sucrose transporter gene (*HbSUT2A*) [23]. In another experiment, the differential expression pattern of isolated cDNA fragments between healthy and TPD-affected trees was confirmed using semi-quantitative RT-PCR. This work revealed the down regulation of *HbTOM20* expression in TPD-affected trees meaning that the gene could play an important role on impaired latex biosynthesis during the occurrence of

TPD [31]. The recent semi-quantitative RT-PCR study showed that a protease inhibitor *PpEPI10* gene was significantly up regulated during colonization of rubber tree by *Phytophthora palmivora* which later interact with the subtilisin proteases in rubber tree. This research revealed the strength of semi-quantitative RT-PCR [32]. In this study, the semi-quantitative RT-PCR on *HbPI* transcripts was able to confirm the relatively stable expression among five conditions in rubber tree clone PB 260. This study has also shown its practical use as the alternative of transcript accumulation analysis for a group of genes in diverse conditions.

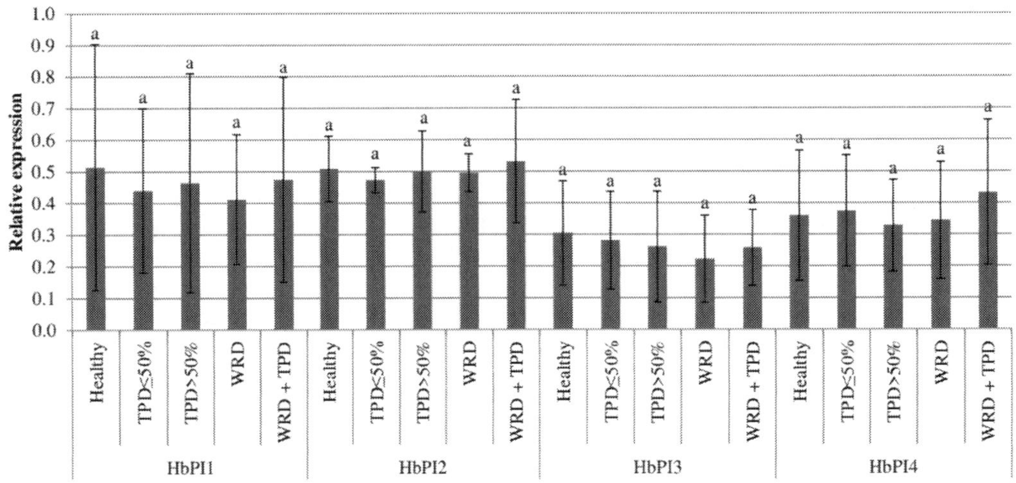

Figure 4. Transcript expression analysis using semi-quantitative RT-PCR for four *Hevea brasiliensis* transcripts encoding protease inhibitors (HbPIs). Each point represents tree biological replicates. Relative expression was calculated as the expression ratio of each treatment and *HbActin*. WRD: white root disease; TPD: tapping panel dryness. Statistical analysis was carried out using Newman-Keuls test of XL-STAT with p-value of 0.05.

3.3. Potential post-transcriptional and post-translational regulation of HbPIs

The stable expression of semi-quantitative RT-PCR analysis throughout different abiotic and biotic stresses highlighted a potential hypothesis of post-transcriptional or/and post-translational regulation of HbPIs. Transcriptional regulation of protease inhibitors has been demonstrated and reviewed [33]. However, the post-transcriptional and post-translational regulation of protease inhibitors in plants has not yet known. Even though so, the biological interaction of protease and its protease inhibitors is certain [34]. The expression of a serine protease inhibitor *KTI4* from tomato was regulated by a protease encoding gene *SlVPE3* in which this protease is post-transcriptionally regulated by miRNA [35]. Other demonstration showed that a cysteine protease inhibitor, cystatin, was transcriptionally regulated by a cysteine protease which was post-translationally regulated by protein modification [36]. The results revealed the complex regulation between protease inhibitors and its target proteases.

3.4. Future perspectives

Finally, the semi-quantitative RT-PCR method was shown to be consistently applicable for transcript expression studies. The work presented in this paper was an attempt to reveal the potential application of rubber tree latex not only as a raw material for tire industry but also for other purposes. The protease inhibitors are among other essential proteins contained inside laticifers. Future studies should be focused on deep functional characterization of *HbPI* gene family to reveal its wide and important role inside the latex.

4. Conclusions

In this study, the semi-quantitative RT-PCR on *HbPI* transcripts was able to confirm the relatively stable expression under abiotic and biotic stresses in rubber tree clone PB 260. The result highlighted the hypothesis of potential post-transcriptional and post-translational regulation of *HbPIs* which defines its abundance in the cell.

5. Acknowledgments

The authors thank the Cooperation and Research Partnership, Assessment, and Agricultural Development (KP4S) program from the Indonesian Ministry of Agriculture for funding the project in 2015. The authors equally thank Siti Maryanti and Niyyah Fitranti for sample preparation.

References

[1] Venkatachalam P, Geetha N, Sangeetha P and Thulaseedharan A 2013 *Afr. J. Biotechnol.* **12** 1297
[2] Mooibroek H and Cornish K 2000 *Appl. Microbiol. Biotechnol.* **53** 355
[3] Clément-Demange A, Priyadarshan P M, Thuy Hoa T T and Venkatachalam P 2007 *Plant Breeding Reviews*: John Wiley & Sons, Inc.) pp 177-283
[4] de Faÿ E and Jacob J L 1989 *Physiology of Rubber Tree Latex*, ed J d'Auzac, *et al.* (Boca Raton, Florida: CRC Press Inc) pp 4-14
[5] Castelblanque L, Balaguer B, Marti C, Rodriguez J J, Orozco M and Vera P 2016 *Plant Physiol.* **172** 1032
[6] d'Auzac J 1989 *Physiology of Rubber Tree Latex*, ed J d'Auzac, *et al.* (Boca Raton, Florida: CRC Press Inc) pp 257-79
[7] Kongsawadworakul P and Chrestin H 2003 *Plant Cell Physiol.* **44** 707
[8] Martin M N 1991 *Plant Physiol.* **95** 469
[9] Jacob JL, d'Auzac J and Prévôt JC 1993 *Clinical Rev. Allergy* **11** 325
[10] Aoki Y, Takahashi S, Takayama D, Ogata Y, Sakurai N, Suzuki H, Asawatreratanakul K, Wititsuwannakul D, Wititsuwannakul R, Shibata D, Koyama T and Nakayama T 2014 *Plant Sci.* **225** 1
[11] Kanokwiroon K, Teanpaisan R, Wititsuwannakul D, Hooper A B and Wititsuwannakul R 2008 *Mycoses* **51** 301
[12] Sritanyarat W, Pearce G, Siems W F, Ryan C A, Wititsuwannakul R and Wititsuwannakul D 2006 *Phytochem.* **67** 1644
[13] Bunyatang O, Chirapongsatonkul N, Bangrak P, Henry R and Churngchow N 2016 *Plant Physiol. Biochem.* **101** 76
[14] Bunyatang O, Chirapongsatonkul N and Churngchow N 2013 *J. Plant. Biochem. Biotechnol.* **22** 185
[15] Putranto R A, Siswanto, Mulyatni A S, Budiani A and Tistama R 2016 *E-Journal Menara Perkebunan* **84** 76
[16] Martiansyah I, Putranto R A and Khumaida N 2017 *E-Journal Menara Perkebunan* **85** 53
[17] Gachon C, Mingam A and Charrier B 2004 *J. Exp. Bot.* **55** 1445
[18] Exner V 2010 *Plant Developmental Biology: Methods and Protocols*, ed L Hennig and C Köhler (Totowa, NJ: Humana Press) pp 275-91
[19] Grosdidier M, Aguayo J, Marçais B and Ioos R 2017 *Plant Pathol.* **66** 359
[20] Choquer M, Boccara M and Vidal-Cros A 2003 *Curr. Gen.* **43** 303
[21] Torp A M, Olesen A, Sten E, Stahl Skov P, Bindslev-Jensen U, Poulsen L K, Bindslev-Jensen C and Andersen S B 2006 *Food Control* **17** 30
[22] Marone M, Mozzetti S, De Ritis D, Pierelli L and Scambia G 2001 *Biol. Procedures Online* **3** 19
[23] Tang C, Qi J, Li H, Zhang C and Wang Y 2007 *J. Biochem. Biophys. Methods* **70** 749

[24] Souza A C M F, Souza D R V, Sanabani S S, Giorgi R R and Bendit I 2009 *Brazilian J. Medical Biol. Res.* **42** 791

[25] Putranto R-A, Herlinawati E, Rio M, Leclercq J, Piyatrakul P, Gohet E, Sanier C, Oktavia F, Pirrello J, Kuswanhadi and Montoro P 2015 *International J. Mol. Sci.* **16** 17885

[26] Chang S, Puryear J and Cairney J 1993 *Plant Mol. Biol.* **11** 98

[27] Die J V and Román B 2012 *J. Exp. Bot.* **63** 6069

[28] Fleige S and Pfaffl M W 2006 *Mol. Aspects Medicine* **27** 126

[29] Vermeulen J, De Preter K, Lefever S, Nuytens J, De Vloed F, Derveaux S, Hellemans J, Speleman F and Vandesompele J 2011 *Nucleic Acids Res.* **39** e63

[30] Kuswanhadi, Leclercq J, Rio M A, Sumarmadji and Montoro P 2007 *Advances in Plant Ethylene Research,* ed A Ramina, *et al.*: Springer Netherlands) pp 31-34

[31] Venkatachalam P, Thulaseedharan A and Raghothama K 2009 *Mol. Biotechnol.* **41** 42

[32] Ekchaweng K, Evangelisti E, Schornack S, Tian M and Churngchow N 2017 *PLoS ONE* **12** e0175795

[33] Koiwa H, Bressan R A and Hasegawa P M 1997 *Trends in Plant Sci.* **2** 379

[34] Rustgi S, Boex-Fontvieille E, Reinbothe C, von Wettstein D and Reinbothe S 2017 *Comm. & Integrative Biol.* e1368599

[35] Wang W, Cai J, Wang P, Tian S and Qin G 2017 *Genome Biol.* **18** 47

[36] Solomon M, Belenghi B, Delledonne M, Menachem E and Levine A 1999 *The Plant Cell Online* **11** 431

IOP Conf. Series: Earth and Environmental Science **183** (2018) 012005 doi:10.1088/1755-1315/183/1/012005

Genetic similarity analysis of *Vanda tricolor* Lindley var. Suavis orchids using OPU 03 and OPU 16 RAPD marker

Tony Haryo Wibisono[1], Suratsih[1], E Yulianti[1], L Sugiyarto[1] and I S Mercuriani[1*]

[1]Biology Education Department, Mathematics and Science Faculty, Yogyakarta State University. Jl. Karangmalang, Sleman, D.I. Yogyakarta 55281, fax. (0274)548203

*Email: ixomerc@uny.ac.id

Abstract. The purpose of this research is to understand the genetic similarity level of *Vanda tricolor* Lindley var. Suavis orchid plants that were regenerated from seeds of the same fruit/pod produced by self-pollination. The seeds were sown on New *Phalaenopsis* (NP) *in vitro* culture medium supplemented with 15 % coconut water (CW). Genetic similarity analysis was performed by PCR-based Random Amplification of Polymorphic DNA (PCR-RAPD) technique. The analysis was carried out at 16 months old regenerated plants after sowing (mas). Seven DNA samples were isolated using a modified Castillo method. It was then used to amplify (PCR) OPU 3 and OPU 16 primers. The electropherogram of PCR result was then converted to binary data and analysed using NTSYSpc software version 2.02. The genetic analysis result showed that the genetic similarity of 7 tested plants ranged between 0.4-0.8.

1. Introduction

Vanda tricolor Lindley var. Suavis has a beautiful colour and fragrant flowers. Two petals and sepals of the flower are white with the brownish-red and purple spot, whereas the lip (labellum) is purple with gradation to pink [1]. *V. tricolor* can be found in several regions in the world, one of these is on Merapi mountain [2]. Exploitation by humans and Merapi eruption are the main cause of the endangered plant. To propagate the plant, people usually planted self-pollinated seeds through *in vitro* culture. Self -pollinated of wild-type plants were done to plants with high genetic character similarity. *Ex situ* conservation of *V. tricolor* orchids can generally be carried out by maintenance in the garden or in the laboratory using *in vitro* culture techniques [3]. Although the use of self-pollination technique is assumed to yield genetically homogenous *Vanda* plants, the similarity level has never been measured.

2. Materials and Methods

The seeds of *V. tricolor* derived from the same fruit of self-aged results 12 bst collection from the Laboratory Plant Tissue Culture Department of Biology Education FMIPA UNY were grown *in vitro*. New *Phalaenopsis* (NP) + coconut water 150 ml.L-1 was used as *in vitro* culture medium. Materials for DNA isolation consisted of Castillo extraction buffer (CTAB 10% + EDTA + Tris-HCl + NaCl), polyvinylpyrrolidone (PVP), chloroform: isoamyl alcohol 24: 1, isopropanol, RNase A solution (Thermo Scientific). Materials for DNA amplification consisted of Go Taq® Green PCR Master Mix

(Promega), OPU-03 primer (5'CTATGCCGAC3 ') and OPU-16 primer (5'ACCTTGCGG3'). Materials for electrophoresis were agarose, loading dye, fluorosafe, buffer Tris Buffer EDTA (TBE) 0.5x and 100 bp marker plus DNA ladder (Vivantis). Equipment used for the manufacture of *in vitro* culture media consisted of analytical scales, 1 L Erlenmeyer, 250 ml measuring cup, hot plate + magnetic stirrer, pH stick, clock bottle and autoclave. The tools used for the *in vitro* planting process consisted of Laminar Air Flow (LAF), tweezers, spatulas, Petri dishes and spirit lights. The equipment used for DNA isolation consisted of mortar and stamper, 15 ml tube, centrifuge machine (Kokusan), pipette, water bath and thermometer. Tools for DNA amplification consisted of thermocycler machine (Eppendorf Mastercycle Hamburg), Eppendorf 0.2 μl tube, micropipette and tips. Tools for electrophoresis were electrophoresis chamber (Mupid e-Xu), parafilm, gel moulds, gel braces and tools for documentation in the form of cameras. UV Vis Spectrophotometer (Genesys) was used to measure the purity of DNA.

The descriptive explorative method was used in this research. The study was conducted in three stages of *in vitro* culture, DNA isolation and DNA amplification by PCR-RAPD technique. *In vitro* cultures were performed to obtain plant samples with subculture and incubation processes. DNA isolation was performed by Castillo method with modification at a 3,500 rpm centrifugation rate [4]. The plant sample was crushed with the addition of 0.1 gram of PVP and 6 ml of Castillo buffer then incubated in a water bath for 30 min at 65°C. The extraction was done by adding 6 ml of CI solution and centrifuged at 3,500 rpm for 20 min. The obtained supernatant was then extracted again. The supernatant on the second extraction added 5 ml of cold isopropanol and incubated at 4 °C. for one night (± 16 hours). The solution is then centrifuged at 3,500 rpm for 20 min. The obtained pellets were then resuspended with aquadest and RNase degradation was performed with the addition of RNase and then incubated at 37°C for one night (± 16 hours). For DNA amplification used composition as much as 1x Go Taq PCR Mix, 10 ng DNA templates, 5 μM primer with total volume 25 μl. The PCR process was carried out at 95°C. for 3 minutes for pre-denaturation followed by 45 repeating cycles consisting of denaturation at 95°C. for 30 seconds, annealing at 27°C for OPU-03 and 36 °C primers for primary OPU -16 for 30 seconds and extension at 70°C for 25 seconds. Final extension is carried out at 70°C for 5 minutes. The collecting of qualitative data of DNA was done by an electrophoresis method. The DNA concentration data were obtained by using Qubit Fluorometer, while for DNA purity was calculated using the UV Vis spectrophotometer at an absorbance of λ260 and λ280 by the formula:

$$\frac{\lambda 260}{\lambda 280}$$

RAPD data analysis is done by scoring on the amplification result. The emerging band is scored 1 while the one that does not appear is given a score of 0. The data is then incorporated into the NTSYS 2.0.2 program to calculate the coefficient of simple matching equations and create a dendrogram using the Unweighted Pair Group Method with Arithmetic Averages (UPGMA) method [5].

3. Results and Discussion

The successful DNA amplification results using OPU 03 and OPU 16 primers was indicated by the appearance of the bands in all samples including three samples of orchid plants *V. tricolor in vivo* used as a benchmark for the identification of genetic uniformity. Band patterns generated from the amplification process with primers OPU 03 and OPU 16 indicated the presence of genetic variation. The amplified DNA band with OPU 03 primer exhibited a monoformic band, while the amplified DNA band with OPU 16 primer showed a polymorphic band (**Figure 1**). Hoon *et al.* (1999) show a high polymorphism band on the amplification results using OPU 03 primer in some species of the genus *Vanda* orchid [6]. This indicates that primer OPU 03 is better at showing polymorphism at the level of interspecies, whereas based on the primary visualization results of OPU 16 it is better to show polymorphisms at the level of intraspecies. The successful amplification of genomic DNA using

RAPD is largely determined by the sequence of primary bases used and their quality or primer content in each reaction.

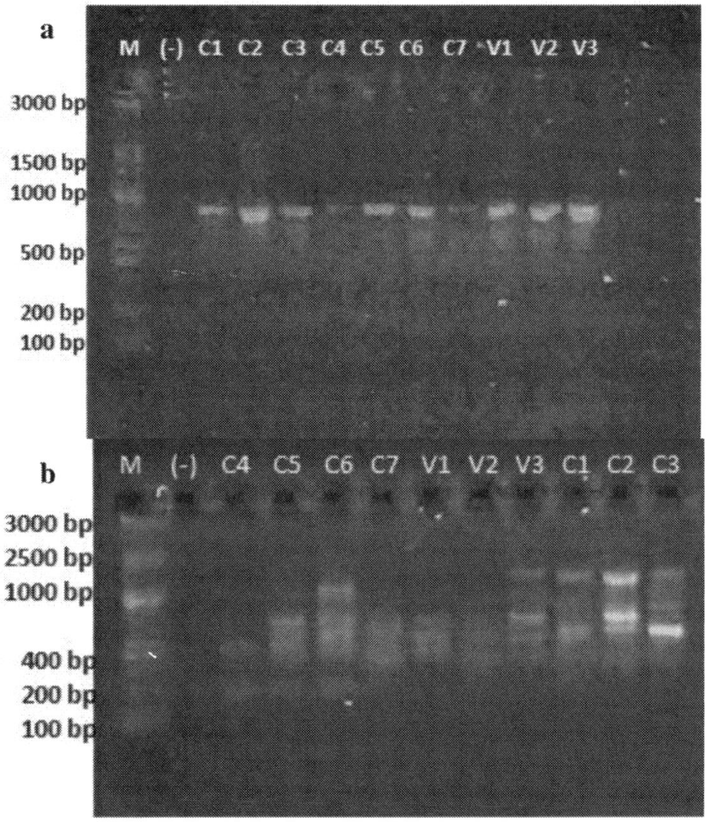

Figure 1. Electropherogram of DNA amplification using primer OPU 03 (a) and OPU 16 (b); (C) *in vitro* plants 2, (C3) plants *in vitro* 3, (C4) plants *in vitro* 4, (C5) *in vitro* 6, (C6) *in vitro* plants 7, (V1) *in vivo* plants 1, (V2) plants *in vivo* 2, (V3) plants *in vivo* 3.

Analysis of genetic uniformity using NTSYSpc 2.0.2 software resulted in the orchid dendrogram of *V. tricolor* (**Figure 2**). The value of genetic similarity coefficient on orchid samples *V. tricolor* ranged between 0.4-0.8. These values indicate that the genetic similarity in the seven plants (C1-C7) has a considerable range. The range of coefficient values of genetic similarity far enough signifies that genetic diversity was present in all seven analysed plants. The genetic diversity that arises can be generated by plants derived from the seeds of non-homozygous pollination. Such diversity can naturally be attributed to Mendelian diversity due to the segregation of both parent genes (heterozygous). Although the RAPD technique used in this study has advantages in terms of ease, it also lacks the ability to identify heterozygotes [7]. The genetic diversity needs to be identified again at the molecular level using a marker other than RAPD capable to identify heterozygotes. *V. tricolor* orchid plant *in vitro* which has the highest similarity level was shown by C1 with C2 and C4 with C7 which has the value of genetic equation coefficient of 0.8. This means that the sample has a genetic similarity rate of 80%. While both sample C2 and C5 have the lowest level of genetic similarity of 0.4. This means that the level of similarity C2 with C5 is only about 40%. The orchid plant *V. tricolor in vivo* used as a comparator is also known to have genetic diversity. If *in vitro* and *in vivo* samples are compared, it can be observed that the highest genetic similarity coefficient was shown by sample C3 with V3 of 0.9, while the smallest distance coefficient value was shown by C6 with V3 of 0.5.

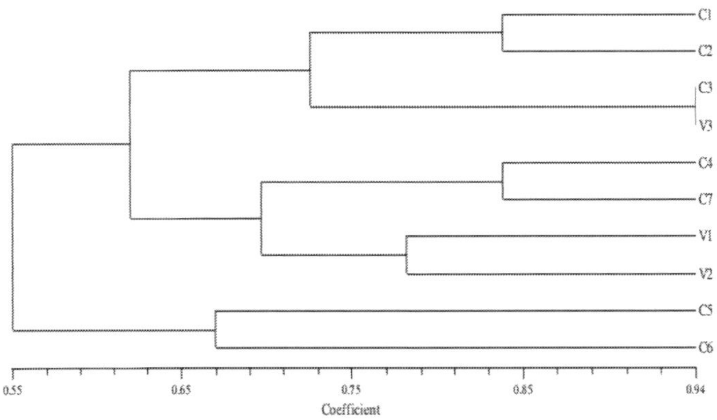

Figure 2. Dendogram of seven orchid plants *V. tricolor* grown *in vitro* and three orchid plants *V. tricolor* grown *in vivo* (C1-C7) *in vitro* samples; V1-V3 sample *in vivo*.

In addition to the internal factors of individual plants caused by heterozygous parent mating, the genetic variation that appears in the seven plants (C1-C7) can be caused by external factors that occur during *in vitro* cultures. Genetic variation that emerged *in vitro* cultures is a constraint in the effort of uniform plant propagation. These constraints arise as a result of the use of chemicals that at certain levels can function as mutagenic such as the use of colchicine and 2,4-D with high concentrations. In addition, the use of inappropriate techniques in *in vitro* cultures allows for variation by either gene mutation or activation of the jump element that allows insertion, deletion and duplication. Variations occur mainly due to repetitive and uncontrolled subcultures resulting in high rates of multiplication [8].

4. Conclusions

Genetic similarity analysis was successfully performed using PCR-based Random Amplification of Polymorphic DNA (PCR-RAPD) technique. Based on the research, it can be concluded that the results analysis of genetic similarity on seven orchid plants *V. tricolor* (C1-C7) self-pollination grown *in vitro* ranged from 0.4-0.8.

5. References

[1] Semiarti, Endang and Rozikin 2015 Prosiding Seminar Nasional Masyarakat Biodiversitas Indonesia **1**(3) (Solo: Central Java) p404
[2] Razak A H 2016 Anggrek *Vanda tricolor* Merapi Langka, Ini Upaya TNGM [internet] available from http://www.harianjogja.com/baca/2016/10/14/anggrek-vanda-tricolor-merapi-langka-ini-upaya-tngm-760619. Accessed at 13 Juli 2017
[3] Dwiyani R, Purwanto A, Indrianto A and Semiarti E 2012 *Jurnal Bumi Lestari* **12** 93
[4] Prahaditya, D 2013 Genetic Diversity Analysis of of turmeric and temulawak plants using *Random Amplified Polymorphic DNA-Polymerase Chain Reaction* (RAPD-PCR) using OPA-OPD 6-10 primers[thesis] (Bogor: Bogor Agricultural University) (In Indonesian)
[5] Riyantini, Indah, Yuniar M and Agung M U K 2014 *J. Akuatika* **5** 63
[6] Hoon L, Saw, Priscilla C T, Yew-Hwa L and Chang-Jin G 1999 *Annals of Botani* **8** 193
[7] Syafaruddin and Nasution M A 2012 *Buletin RISTRI* **3**(2) 177
[8] Santoso U and Nursandi F 2003 *Kultur Jaringan Tanaman* (Malang: UMM Press)

IOP Conf. Series: Earth and Environmental Science **183** (2018) 012006 doi:10.1088/1755-1315/183/1/012006

Microscopic callus selection of sengon tree (*Falcataria moluccana*) putative tolerant to *Uromycladium falcatarium*

Asri Insiana Putri[1]*

1Centre for Forest Biotechnology and Tree Improvement, Jalan Palagan Tentara Pelajar KM 15, Purwobinangun, Pakem, Sleman, Yogyakarta 55582, Indonesia

*E-mail: asriip@yahoo.co.id

Abstract. The increase of the cultivation area of the sengon tree (*Falcataria moluccana*) through monoculture cropping system especially in Java has caused an increase in gall rust disease. Infestation of the disease may lead to 100% death of sengon seedlings. *Uromycladium falcatarium* has been identified as the fungus causing the gall rust epidemic. This research aimed to select calli putative tolerant to gall rust disease. Seeds of ten individual colonies of 3 clones of sengon trees from Purwobinangun district that have been attacked severely by gall rust were used as explant sources. Gall rust filtrate was used as bioagent selection *in vitro* and the viable calli were selected under a microscope. The callus viability was determined based on the average living cell percentage stained with *fluorescence diacetic acids*. These viable cells of callus were found to have a good correlation with the tolerance index value. The results showed that callus colony from clone 3 individual 4, clone 5 individual 5, and clone 6 individual 5 exhibited the highest tolerance index value. The cell with different index values to the biotic pathogen filtrate indicated the presence of an enzymatic defence process of the cells against gall rust attack.

1. Introduction

The increasing cultivation of sengon tree (*F. moluccana*) in the Java island using monoculture cropping system leads to the increase of gall rust disease. This disease may cause 100% death of the sengon seedlings. *Uromycladium falcatarium* sp. Nov. has been identified as the fungus that causes the gall rust epidemic in South East Asia [1]. Due to its obligate parasite, this fungal pathogen is the major reason for the tree product loss. The disease not only reduces the yield but also greatly impair the quality and stability of production, undermining efforts to promote sustainable forestry. Environmental and health hazards resulting from the application of numerous chemical fungicides have raised concerns.

One of the forest-breeding programmes is to develop tolerant and/or resistant clones of valuable woody trees. To enhance the efficiency and success rate of this breeding programme, more rapid and cost-effective screening methods are needed. Mass screening methods can be performed with well-specified filtrates or purified toxins applied to the callus organs. Purified toxins are more easily quantified and replicated than inoculum and crude extracts (filtrates). There are many pathogens from which toxins have not yet been isolated (mostly obligate biotrophic parasites). However, before using a toxin as a selection agent for disease resistance, it is important to determine whether the toxin is an essential component of the pathogenicity and disease development [2]. The use of crude extracts or

IOP Conf. Series: Earth and Environmental Science **183** (2018) 012006 doi:10.1088/1755-1315/183/1/012006

filtrates on callus as explants are more laborious than the use of toxins. Nevertheless, it may be the better choice for plant-pathogen interactions using unknown toxins such as the gall rust disease.

In vitro selections may also be difficult and time-consuming. In many papers published since 1980, many of the problems related to the theoretical and practical approaches of *in vitro* selections and their importance for plant breeding have been addressed [3, 4, 5, 6, 7, 8].

This research provides an overview of the basic principles and methodology of *in vitro* callus selections for rust gall disease resistance in sengon tree. The gall rust filtrate of sengon tree contains toxic secondary metabolites such as alkaloids, steroids, saponins and terpenoids. These metabolites belong to physianticipin group, low molecular weight compounds produced constitutively, in contrast to the phytochemical groups synthesized in the course of pathogen infection. The active compounds were suspected derived from sengon tree as host plant [9]. The objectives of this research were microscopic callus selection of *F. moluccana* putative tolerant to *U. falcatarium* sp. Nov.

2. Materials and Methods

This research has been conducted at the Tissue Culture Laboratory, Centre for Forest Biotechnology and Tree Improvement, Yogyakarta. *In vitro* method by Wagh *et al.* [10] was used with a modification in this research. Calli were induced from seeds of sengon trees from Purwobinangun district, Yogyakarta. Liquid MS [11] media with callus initiation hormone added with gall rust tumour filtrate were used for cell selection. Gall rust filtrate (25% *dimethyl sulfoxide* solution) was used as a chemical agent for callus *in vitro* selection. The observation of cell viability was using the FDA (*fluorescence diacetic acid*) live cell painting method [12]. Callus cell selection of sengon *in vitro* was measured based on the highest index values obtained from several factors where 30 day observation period divided by mean callus induction time on explants, an incubation time of LD_{50} callus infiltrate gall rust treatment, the percentage of explants that form a callus, and living cell percentage on FDA cell paint. Flowchart of microscopic callus selection of *F. moluccana* putative tolerant to *U. falcatarium* is shown in **Figure 1**.

Individual clone SPW-3, clone SPW-5, and clone SPW-6

⇩

Explants from seeds

⇩

Callus induction

⇩

Multiplication of callus

⇩

Immersion of callus colony sample individually in gall rust filtrate with LD_{50} incubation time treatments

⇩

Callus viability observation using FDA (*fluorescence diacetic acid*) based on living cells at 3 microscopic views in incubation time of LD_{50}

⇩

Regeneration of colonies of selected sengon clones based on the highest index value

Figure 1. Flow chart of microscopic callus selection of *F. moluccana* putative tolerant to *U. falcatarium*.

IOP Conf. Series: Earth and Environmental Science **183** (2018) 012006 doi:10.1088/1755-1315/183/1/012006

3. Results and Discussion

The results showed that callus cells of sengon tree of clone 3 individual 4 (**Figure 2**), clone 5 individual 5 (**Figure 3**), and clone 6 individual 5 (**Figure 4**) had the highest tolerance index values.

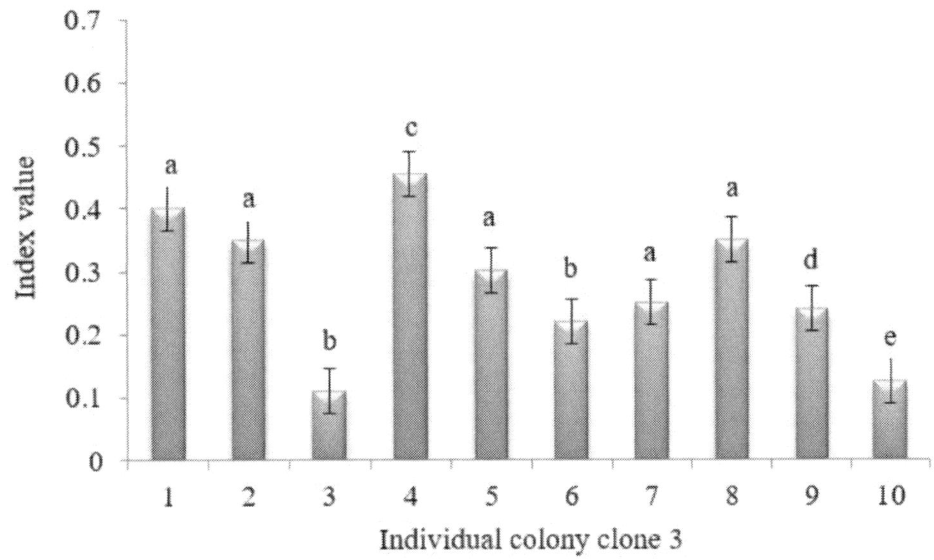

Figure 2. The index value of individual sengon clone 3.

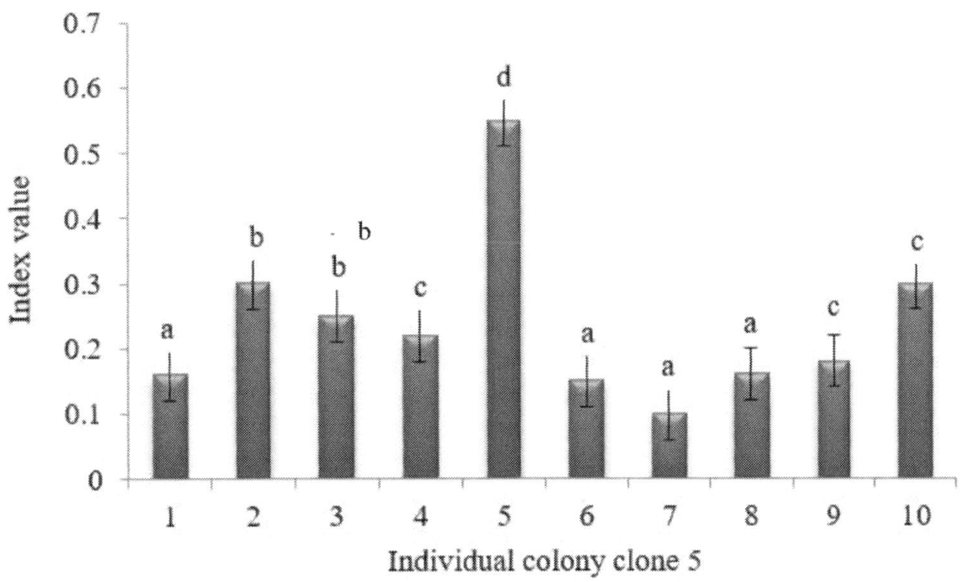

Figure 3. Index value of individual sengon clone 5.

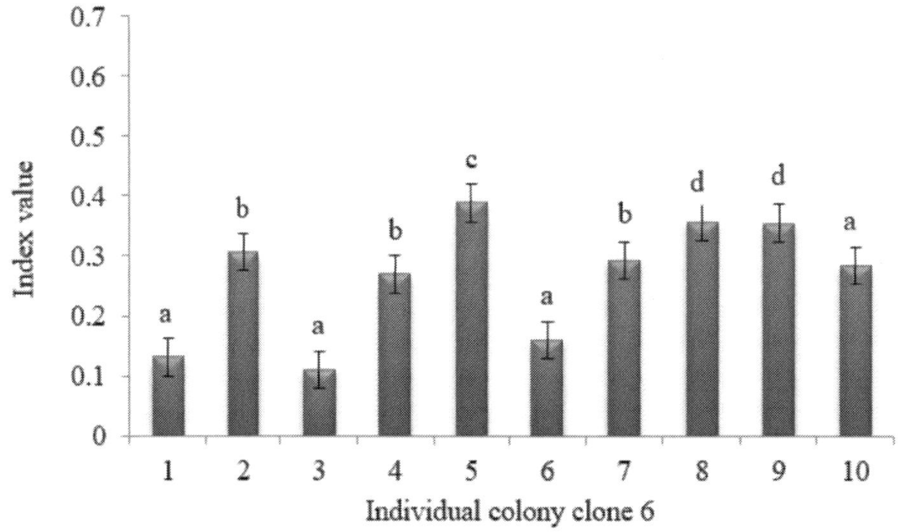

Figure 4. Index value of individual sengon clone 6.

Callus colonies from the 3 clones grew well *in vitro*. Callus clone selections of sengon tree candidate tolerant to gall rust disease are shown in Figure 5 and its regeneration after callus selection is shown in **Figure 5**.

There are different approaches to acquire genotype selections through the breeding of tolerant trees, among others are indirect selections *in vitro*. However, until now there has not be known yet exactly what parameters should be used. Index values of callus cells of sengon *in vitro* based on the day observation period, mean callus induction time on explants, an incubation time of LD_{50} callus in filtrate gall rust treatment, the percentage of explants form callus and viable cell percentage should be a good alternative for selection *in vitro*. The individual family of SPW 6-5 colony callus with the highest index value gave the best growth compared to an individual family of SPW 6-7 colony callus with the lowest index value (**Figure 6**).

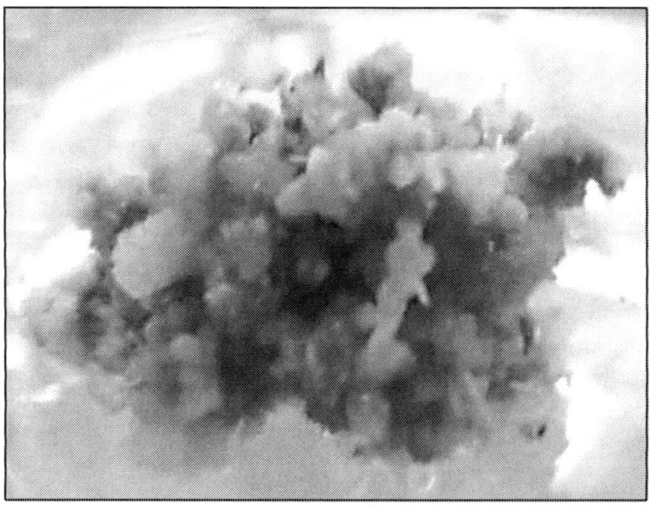

Figure 5. Regeneration of sengon callus clone SPW-6.5 after 42 days in culture.

IOP Conf. Series: Earth and Environmental Science **183** (2018) 012006 doi:10.1088/1755-1315/183/1/012006

Figure 6. Callus selection of sengon candidate gall rust tolerant: gall rust attacked at sengon trees as seeds source (A), suffering from gall rust attacks (A.1); plantlets of sengon tree (B); callus initiation of sengon *in vitro* (C); callus immersion in gall rust filtrate (D); individual clone SPW-6.5 had the highest index value (E); individual clone SPW 6.7 had the lowest index value (F).

In vitro selection method has been highly more effective and efficient [5, 13, 14, 15, 16] compared with the tolerance selection directly at the nursery or the selection process in the wild. In this research, good callus colonies that regenerated fast, friable with shiny white and green were obtained only in 42 days of *in vitro* culture. The maintenance of callus was important for the subsequent research until the seedlings were regenerated from the plantlets. This research was conducted to avoid a lack in developing a woody plant tolerant to a certain disease using conventional ways i.e. the low accuracy selection, the lengthy time needed to develop and to test the progenies of the previous mother plants.

On the indirect selection *in vitro*, the cell viability indicated with *fluorescence diacetic acids* has a good correlation with the tolerance index value. The difference in the cell reaction on the index value parameter of the biotic pathogen filtrate indicates the presence of an enzymatic defence process of the sengon callus cell. Kumar *et al.* [17] reported that the pathogen filtrate induced oxidative pressure during the *in vitro* selection. During the agent selection treatment, the cells reproduced a radical superoxide which was immensely toxic to protect the cells from the pathogen filtrate pressure.

The presence of steroids, saponins, and terpenoids in the gall rust filtrates was considered as the agent selection compounds *in vitro* that contribute to the sengon tree defence mechanism [9]. Therefore, sengon cells putative tolerant to gall rust filtrate *in vitro* that survived from oxidation damage because the cells had high anti-oxidative enzyme activity. Programmed cell destruction (PCD) is a cellular homeostatic reaction as a defensive response against biotic pathogen stress or biochemical gall rust filtrate selection agent. As it reaches closer to a unicellular level, the cell selection process has been narrowing down even further as the homeostatic reaction became more potent, this was able to stabilize smaller internal cells [18].

4. Conclusions

The difference in the cell reaction based on the index value parameter to the biotic pathogen filtrate from microscopic callus selection indicates the presence of an enzymatic defence physiology process in callus cells of *Falcataria moluccana* putative tolerant to *Uromycladium falcatarium*.

5. Acknowledgements

The author is grateful to the Centre for Forest Biotechnology and Tree Improvement in Yogyakarta for the financial support.

6. References

[1] Doungsa C, Mc Tagart A and Geering A D W 2015 *Australasian Plant Pathol.* **44** 25
[2] Joint FAO 2010 Mass Screening Techniques for Selecting Crops Resistant to Diseases. Joint FAO/IAEA Programme of Nuclear Techniques in Food and Agriculture. International Atomic Energy Agency Vienna (Vienna: Italy)
[3] Shepard J F 1981 *Ann. Rev. Phytopathol.* **19** 145
[4] Wenzel G, Foroughi-Wehr B, Friedt W, Köhler F and Oo T 1985 *Hereditas* **103** 15
[5] Daub M E 1986 *Ann. Rev. Phytopathol.* **24** 159
[6] Buiatti M and Ingram D S 1991 *Experientia* **47** 811
[7] Graniti A 1991 *Experientia* **47** 751
[8] Svábová L and Lebeda A 2005 *J. Phytopathol.* **153** 52
[9] Putri A I, Na'iem M, Sapto I and Rahayu S 2016 *Falcataria moluccana* Genetic Selection of *Uromycladium falcatarium* Putative Tolerant [Dissertation] Gadjah Mada Univ. Yogyakarta: Indonesia
[10] Wagh G S, Chavhan D M, Sayyed M R G 2013 *J. Environ. Res. Technol.* **3** 93
[11] Murashige T and Skoog F 1962 *Physiol. Plant.* **15** 473
[12] Jones A M P A, Chattopadhyay A, Mukund S, Zoń J, Saxena P K 2012 *Biomed. Central Plant Biol.* **12** 75
[13] Ingram D S and Helgeson J P 1980 *Tissue Culture Methods for Plant Pathologists* (Oxford: Blackwell Scientific Publications) 272 p
[14] Helgeson J P and Deverall B J 1983 *Use of Tissue Culture and Protoplasts in Plant Pathology* (Sidney: Academic Press) 194 p
[15] Huang J S 2001 *Plant Pathogenesis and Resistance - Biochemistry and Physiology of Plant-Microbe Interactions* (Dordrecht: Kluwer Academic Publishers) 691 p
[16] Lebeda and Švábová L 2010 *In vitro* Screening Methods for Assessing Plant Disease Resistance. *International Atomic Energy Agency Vienna* (Austria: FAO/IAEA Division of Nuclear Techniques in Food and Agriculture)
[17] Kumar S, Negi S P and Kanwar J K 2008 *In Vitro Cell. Dev. Biol. Plant* **44** 474
[18] Watanabe N and Lam E 2008 *J. Biol. Chem.* **283** 3200

IOP Conf. Series: Earth and Environmental Science **183** (2018) 012007 doi:10.1088/1755-1315/183/1/012007

Somatic embryogenesis of sago palm (*Metroxylon sagu* Rottb.) through suspension culture technique

Imron Riyadi[1,2]*, D Efendi[1], B S Purwoko[1] and D Santoso[2]

[1]Department of Agronomy and Horticulture, Bogor Agricultural University, Kampus IPB Dramaga, Bogor 16680, Indonesia
[2]Indonesian Research Institute for Biotechnology and Bioindustry, Jalan Taman Kencana No.1, Bogor 16128, Indonesia

*E-mail: imron.riyadi@iribb.org

Abstract. Mass production of sago palm (*Metroxylon sagu* Rottb.) through *in vitro* culture has not been fully explored. Suspension culture has a potency for scaling up the production of somatic embryos to support the mass production of the elite clones. This research aimed to determine the composition of cytokinins that efficiently induce somatic embryogenesis of sago palm in suspension cultures. A clump of friable calli derived from apical meristem culture was used as starting material. The calli were cultured on a modified MS medium with 1-5 mg/L kinetin and 1 mg/L TDZ. The results showed that the calli of sago palm grew well. The highest volume of calli at 5.67 mL was achieved at 1 mg/L kinetin + 1 mg/L TDZ. The lowest calli yield at 3.65 mL was produced at 5 mg/L kinetin + 1 mg/L TDZ. The highest fresh weight of sago palm biomass was achieved at 1 mg/L kinetin + 1 mg/L TDZ (4.2 g/flask). The fastest somatic embryo induction was at 7 weeks after culture in the best treatment (1 mg/L kinetin + 1 mg/L TDZ). At the medium supplemented with 1 mg/L kinetin + 1 mg/L TDZ produced on average 133.7 somatic embryos per flask.

1. Introduction

Sago palm (*Metroxylon sagu* Rottb.) is one of the potential crops for supporting food security and bioenergy in Southeast Asia particularly in Indonesia. Sago palms grow well in Eastern parts of Indonesia such as the Moluccas and Papua [1-3]. The area of sago palm in the world was estimated at 2.47 million ha in 1997 [4] then increased became 5.2 million ha in 2013 [3] and more than half grown in Indonesia. Sago starch has been used as a staple food especially in Eastern Indonesia for centuries [4].

Carbohydrate of sago palm could be used as raw materials for various industries including bioethanol [2, 5]. In the food industry, sago starch is a primary ingredient for noodles, cakes, and other food [3]. The starch can also be utilised as raw materials of adhesive gel, pharmaceutical, cosmetics, bioplastics, and paper [4, 6]. Although its industrial importance, large-scale sago palm plantations have not been developed yet. One of the problems was the unavailability of a large number of planting materials. Having proven well in some crops, a technique for somatic embryogenesis of sago palm has not been explored yet to produce mass clonal sago planting materials.

Somatic embryogenesis through liquid cultures have been developed for several plants i.e. yam (*Dioscorea alata* L.) 'Pacala Duclos' [7], black gram (*Vigna mungo* L.) Hepper [8], cucumber

(*Cucumis sativus* L.) [9], *Stevia rebaudiana* [10], and *Wasabia japonica* [11]. Liquid culture is more efficient in using media, labour, and space than solid culture [11]. The liquid culture technique is implemented particularly for automatization, scaling up, increasing growth as well as uniformity of the cultured embryos [12, 13].

Somatic embryogenesis of some monocot plants has been well developed [14]. The uses of somatic embryogenesis techniques to produce clonal planting materials were reported for oil palm [15], coconut [16], date palm [17], peach palm [18] and sago palm [19]. The somatic embryogenesis of sago palm reported previously was utilized solid media culture. In liquid media culture technique, the composition of culture media especially the use of cytokinins played a vital role in somatic embryogenesis [20]. The aim of this research was to determine the composition of cytokinin to effectively induce somatic embryogenesis of sago palm in suspension culture.

2. Materials and Methods

2.1. Plant materials

The research was mostly conducted at the Laboratory of Plant Cell Culture and Micropropagation, the Indonesian Research Institute for Biotechnology and Bioindustry (IRIBB), Bogor, Indonesia. Histological analysis and observation were done at the Microtechnique Laboratory, Faculty of Biology - Bogor Agricultural University, and Research Centre for Biology - The Indonesian Institute of Sciences.

Friable calli were used as starting materials in this research (**Figure 1A**). The calli were initiated from young suckers of sago palm variety of Alitir originated from Merauke, Papua, Indonesia. The calli had been cultured several times before being used.

2.2. Media, treatment and culture conditions

The explants were cultured in a modified liquid MS [21] medium (MMS) as described by Tahardi [19]. The treatments were concentrations of kinetin: 0, 1, 3 and 5 mg/L, and each combined with 1 mg/L TDZ. The volume of the medium was adjusted to 30 ml/100 mL Erlenmeyer flask. The pH of all media was adjusted to 5.7 before being autoclaved at 121 ^{0}C and 1.0 kg/cm^2 for 20 min.

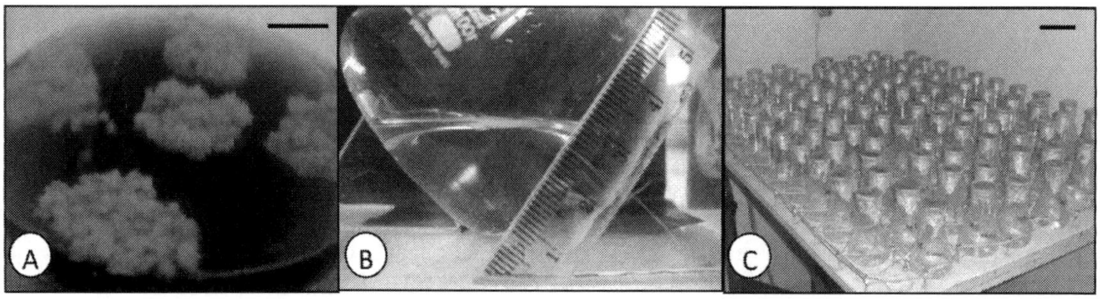

Figure 1. Initial culture process of sago palm: (A) Friable calli as explant source; (B) Initial biomass of sago suspension culture measured by CVS; (C) Sago suspension cultures on an orbital shaker.

All cultures were placed on an orbital shaker at 80 rpm (**Figure 1C**) in the culture room at the temperature of 25 $\pm$ 1 ^{0}C under cool-white fluorescent lamps providing approximately 10 µmol photon/m^2/s over a 12-h photoperiod for 10 weeks with one subculture.

2.3. Biomass growth and somatic embryo induction

Biomass growth was observed using CVS (Cell Volume after Sedimentation) method particularly to determine the growth trend of each treatment and the best growth. Measurement of CVS was done every week using the CVS tool (**Figure 1B**). Fresh weight of the biomass was observed at initial culture and when harvested at the final culture using a digital balance in the laminar air flow hood. All cultures were observed the predominant colour after sieving process and transferred to the Petri dish. Observations were done in the laminar air flow hood. The number of somatic embryos was counted after sieving process and transferred on the Petri dish.

2.4. Histological study

Samples of somatic embryos were collected at the phase of elongated shape. The samples were fixed in 50% formaldehyde/acetic acid/alcohol (FAA) (v/v) The experiment was arranged using a completely randomized design with 10 replications. Data were subjected to one-way analysis of variance (ANOVA). Differences among treatment means were determined by Duncan's multiple range test at $P < 0.05$. To improve the accuracy of the analysis, SPSS version 19 was used as a tool program for data analysis in accordance with the protocol of Johansen [22]. The samples were then transferred into 70% ethanol and dehydrated in an ethanol series. The samples were embedded in paraffin at 60 °C. Longitudinal sections were cut using a rotary microtome to 10 μm thickness. The sections were placed onto slides and stained with Johansen's Safranin and Fast Green for 10 min [23]. The permanent slides were mounted in synthetic resin (Entelan®). Images were captured with an Optilab software camera attached to a Nikon Eclipse E100 microscope. The scales were projected in the same optical conditions.

2.5. Data analysis

The experiment was arranged using a completely randomized design with 10 replications. Data were subjected to one-way analysis of variance (ANOVA). Differences among treatment means were determined by Duncan's multiple range test at $P < 0.05$. To improve the accuracy of the analysis, SPSS version 19 was used as a tool program for data analysis.

3. Results and Discussion

3.1. Biomass growth

The volume of sago palm biomass increased steadily until 10 weeks of culture (**Figure 2**). Fast increases in biomass started from three weeks, reached the peak at six weeks of cultures, and started to decrease afterwards (**Figure 2**). The fastest biomass growth was achieved at 1 mg/L kinetin + 1 mg/L TDZ, whereas the lowest was observed at 5 mg/L kinetin + 1 mg/L TDZ. At the end of 10 weeks of culture, the highest biomass volume at 5.67 mL/flask was achieved at 1 mg/L kinetin in combination with 1 mg/L TDZ, followed by 3 mg/L kinetin + 1 mg/L TDZ (5.29 mL/flask). Whereas the lowest biomass volume was 3.65 mL/flask observed at 5 mg/L kinetin in combination with 1 mg/L TDZ (**Figure 2 and 3**).

Within the first 3 weeks of culture, the increases of sago palm biomass volume were almost the same for all treatments at 0.9 to 1.7 mL/flask. The exponential growth rate of sago palm biomass occurred during the fourth to the sixth weeks of culture in the media supplemented with 1 mg/L kinetin plus 1 mg/L TDZ. These results clearly showed that the composition of the cytokinins was crucial in the biomass growth. Combination of 1 mg/L kinetin and 1 mg/L TDZ was the best among others.

After six weeks in the cultures, the sago palm biomass was subcultured to replenish the corresponding media compositions, marked as culture-2. Compared to the previous culture, the biomass growth rates of cultures-2 were much lower. In this case, the sources of energy in cultures-2 had been shared among the cells more for somatic embryo induction.

In general, the second suspension culture of sago palm showed that the biomass increases at all treatments were lower and almost the same and not significantly different. Therefore, the highest growth of sago suspension in the cultures-2 was achieved persistently in media with 1 mg/L kinetin in combination with 1 mg/L TDZ (**Figure 3**). The lowest biomass growth still occurred in media with 5 mg/L kinetin in combination with 1 mg/L TDZ (**Figure 3**). **Figure 2**, shows the trend of growth and development of sago palm suspension. The growth rate of initial culture (cultures-1 during the first six weeks) was faster than during subculture (cultures-2). After the sago palm suspension has accumulated cellular biomass, the cells developed into aggregates for somatic embryo induction (developmental step). Cytokinin combination between 1 mg/L kinetin and 1 mg/L TDZ hastened somatic embryogenesis. Some papers reported that kinetin and TDZ application as single or in combination with other cytokinins induced somatic embryogenesis on plants like *Limonium sinensis* using 4.54 µM TDZ [24], *Coffea canephora* P ex Fr. using 0.93 µM kinetin [25], *Echinacea purpurea* L. using 2.5 µM TDZ [20]. Similar to that of Cinchona [26], the growth of sago palm biomass in suspension culture increased initially. Furthermore, when induction and maturation phases of somatic embryos occurred, it decreased the growth rate of biomass on sago palm [27].

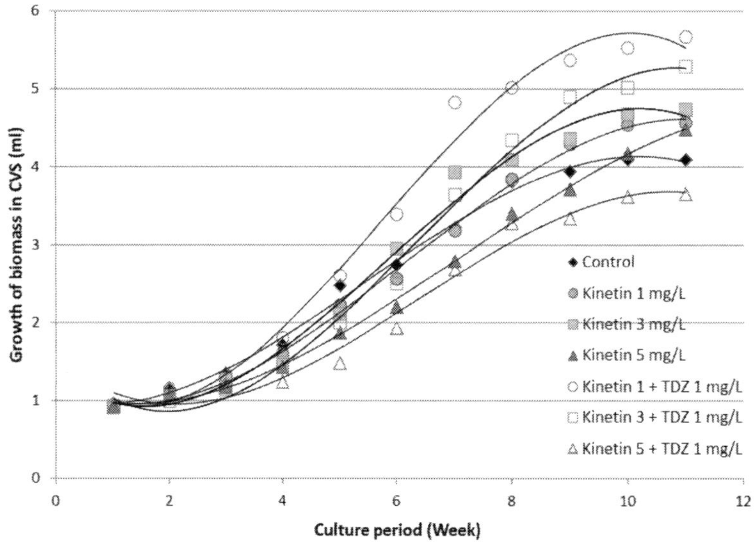

Figure 2. Biomass growth curve of sago palm suspension in CVS from the initial culture until harvesting time at ten weeks.

Table 1 shows the fresh weights for all treatments at harvest. The highest fresh weight was 4.164 g/flask achieved at media with 1 mg/L kinetin in combination with 1 mg/L TDZ that appropriate with the achievement of highest biomass in CVS. Fresh weight was not always the same as CVS of sago palm biomass. Treatment media with 5 mg/L kinetin in combination with 1 mg/L TDZ resulted in the lowest biomass CVS (3.65 ml/flask) but it resulted a fresh weight higher than control treatment. The lowest fresh weight was 3.128 g/flask obtained at control treatment, whereas at media with 5 mg/L kinetin in combination with 1 mg/L TDZ resulted 3.739 g/flask although its biomass CVS was lower (3.65 mL/flask) than control treatment (4.09 mL/flask) (**Table 1**).

3.2. Predominant colour

In the initial culture, the callus was yellowish (**Figure 1A**). Furthermore, along with the growth and development of sago palm cultures, changes in colour, shape and size occurred. The colour change from yellowish to whitish occurred after four weeks of culture (culture-1). The callus was bigger and became more globular. The subculture stage (culture-2), produced the bigger callus with shape and colour clearly identified as proembryoids. In ten weeks when sago palm suspension was harvested, somatic embryos produced had developed to globular, elongated and scutellar shaped (**Figure 6A-D**).

The predominant light-yellow colour of somatic embryos occurred in almost all treatments at harvesting time. Somatic embryos which were brownish-yellow and bigger appeared in most treatments media with 3 mg/L kinetin in combination with 1 mg/L TDZ and media with 5 mg/L kinetin in combination with 1 mg/L TDZ (**Table 1**). Change in colour of somatic embryos from yellowish to reddish in the maturation stage occurred on solid media [28].

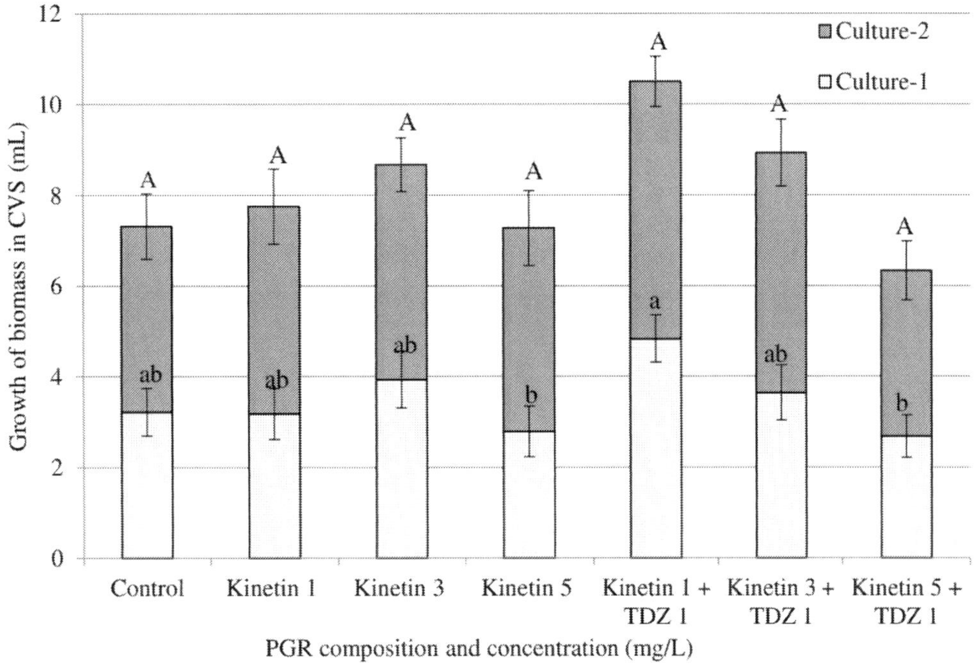

*)The same bar with different letters (capital and lowercase) are significantly different according to Duncan's multiple range test at $P < 0.05$.

Figure 3. Biomass growth of sago palm suspension cultures in CVS in the first culture and the second culture (subculture).

Table 1. Fresh weight and predominant colour of cultures when harvesting at ten weeks.

PGR's Concentration (mg/L)		Fresh weight (g/flask)	Predominant colour
Kinetin	TDZ		
0	0	3.128 ± 0.402	Light-yellow
1	0	3.234 ± 0.736	Whitish-yellow
3	0	3.737 ± 0.736	Light-yellow
5	0	3.736 ± 0.572	Light-yellow
1	1	4.164 ± 0.221	Light-yellow
3	1	3.796 ± 0.247	Brownish-yellow
5	1	3.739 ± 0.434	Brownish-yellow

Values are mean ± SE from 10 replicates per treatment. Initial fresh weight: 0.3 g/flask

3.3. Somatic embryo induction

Sago palm callus suspension cultures started to form somatic embryos at 7 to 9 weeks after culture. Some of the white calli changed to mostly white yellowish or greenish with solid structures called globular somatic embryos when somatic embryogenesis was initiated clearly in the Erlenmeyer flasks

(**Figure 6A**). The concentration of kinetin at 1 mg/L combined with 1 mg/L TDZ accelerated the initiation of somatic embryogenesis. All cultures formed somatic embryos at nine weeks after culture (**Figure 4**).

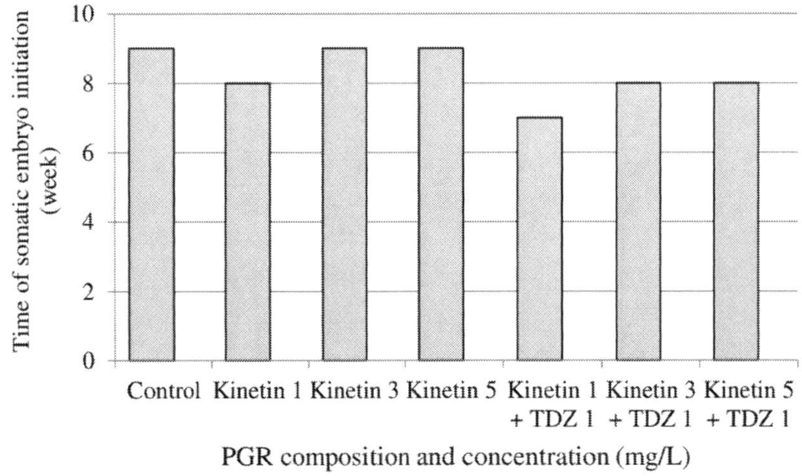

Figure 4. Average of time for somatic embryo initiation of sago palm suspension culture.

The highest number of somatic embryos induced (133.7 pcs) was achieved at media with 1 mg/L kinetin plus 1 mg/L TDZ and it was significantly different from K5 and the control but was not significantly different from media with 1 and 3 mg/L kinetin (as single), 3 mg/L kinetin + 1 mg/L TDZ and 5 mg/L kinetin + 1 mg/L TDZ. Whereas the lowest somatic embryo induction was obtained at control treatment (39.9 pcs). There were more mature somatic embryos obtained in the treatment combination between kinetin with TDZ than that of kinetin as a single treatment (**Figure 5**).

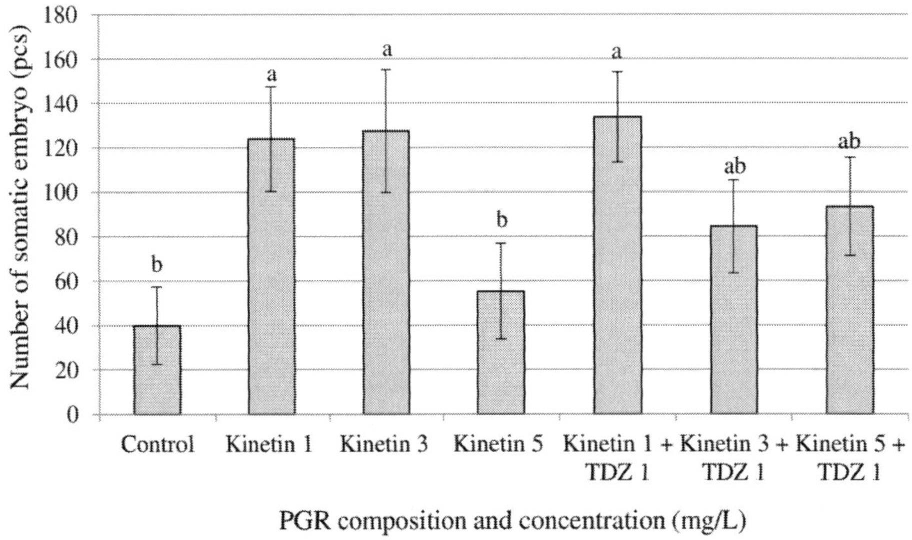

*)Bars with different letters are significantly different according to Duncan's multiple range test at $P < 0.05$.

Figure 5. Number of somatic embryos induced on all treatments at ten weeks after culture.

IOP Conf. Series: Earth and Environmental Science **183** (2018) 012007 doi:10.1088/1755-1315/183/1/012007

Somatic embryos were induced at all treatments. There were differences in time of somatic embryo formation and number of somatic embryos. Somatic embryos which were inducted from culture-1 during the six weeks cultures (older somatic embryos) were slightly different from somatic embryos from culture-2 (new somatic embryos) at the ten weeks. The differences in size (**Figure 6C and 6D**) and colour of the somatic embryos. The new somatic embryos were whitish and yellowish-white and the older somatic embryos were reddish, light-yellow or brownish-yellow (**Figure 6B**).

Addition of cytokinin increased somatic embryo induction whether as single PGR or in combination. Kinetin affected induction of somatic embryo when used as single treatment. Combination of kinetin and TDZ in proper concentrations increased the induction of somatic embryos [20, 29, 30]. This was confirmed also in sago palm cultures. In this study, it was found that 1 mg/L kinetin combined with 1 mg/L TDZ promoted the highest frequency of somatic embryo induction.

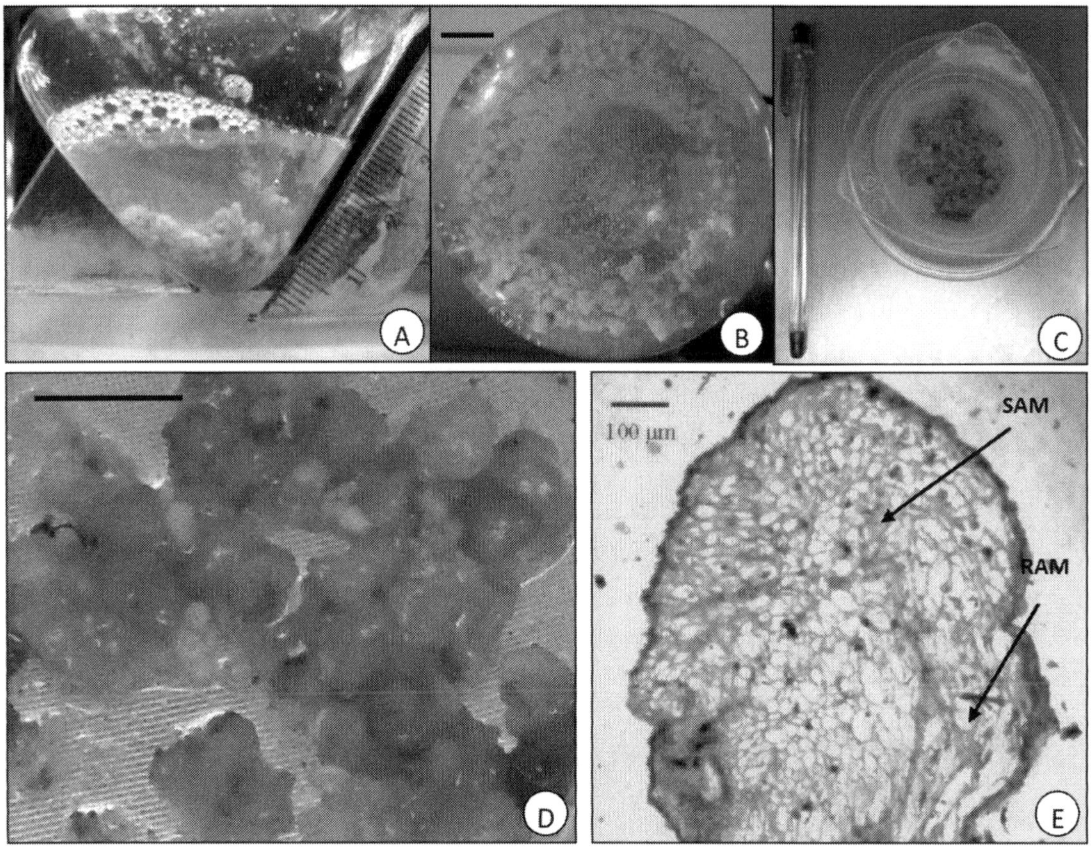

Figure 6. Somatic embryo of sago palm in suspension culture at 10 weeks after culture: (A) Biomass volume measurement by CVS; (B) Aggregate of somatic embryos; (C) Sieving for somatic embryo aggregates from suspension culture; (D) Somatic embryos after sieving; (E) Longitudinal section of somatic embryo at elongated stage. Scale bar = 1 cm (B and D). Length of a pen = 15.5 cm (C). **SAM**: Shoot Apical Meristem, **RAM**: Root Apical Meristem.

3.4. Histological study

The result of the histological analysis showed that cell structure was dense and homogeneous. The histological analysis displayed bipolar structures that comprised the top (shoot apical meristem, SAM) and bottom (root apical meristem, RAM) sides [31].

The top side that contains SAM was the shoot primordium and the bottom side that contains RAM was the root primordium (**Figure 6E**). A well-formed and dense structure indicated normal and good somatic embryo with strong potency to develop into new individual plants [32]. The existence of SAM and RAM in somatic embryos will support the germination process of somatic embryos became normal plantlets having shoot and roots [33].

4. Conclusions

Somatic embryos of sago palm can be induced in a suspension culture. The best composition medium was 1 mg/L kinetin combined with 1 mg/L TDZ which resulted in the highest of biomass volume (5.67 mL/flask), biomass fresh weight (4.164 g/flask), and somatic embryo number (133.7 pcs/flask). During somatic embryo induction, the colour of calli has changed from mostly yellowish to light-yellow, whitish-yellow and brownish-yellow.

5. Acknowledgements

The authors thank Riset Perkebunan Nusantara (RPN), Bogor, Indonesia for providing a grant to Project of Sago Palm through Revitalization Program.

6. References

[1] Abbas B, Bintoro M H, Sudarsono, Surahman M and Ehara H 2009 *Biodiversitas* **10** 168
[2] Ehara H 2009 *J. Agron. Indonesia* **37** 209
[3] Bintoro H M D. Amarilis S S. Dewi R K and Ahyuni D 2013 *Sagu, mutiara hijau khatulistiwa yang dilupakan.* (Digreat Publishing) p 113. (*in Indonesian*)
[4] Flach M 1997 *Sago Palm Metroxylon sagu Rottb.* International Plant Genetic Resources Institute, Rome-Italy p 76
[5] Kanro M Z, Rouw A, Widjono A, Syamsuddin, Amisnaipa and Atekan 2003 *J. Litbang Pertanian* **22** 116
[6] Rostiwati T, Lisnawati Y, Bustomi S, *et al.* 2008 *Sagu (Metroxylon spp.) sebagai Sumber Energi Bioetanol Potensial.* Badan Litbang Kehutanan dan Perkebunan, Jakarta p 85
[7] Jova M N, Rafael G K and Ernesto E C 2011 *Biotechnol. Agron. Soc. Envi.* **15** 515
[8] Muruganantham M, Amutha S and Ganapathi A 2010 *In Vitro Cell. Dev. Biol. Plant* **46** 34
[9] Konstas J and Kintzios S 2003 *Plant. Cell. Rep.* **21** 538
[10] Gupta P, Sharma S and Saxena S 2014 *Appl. Biochem. Biotechnol.* **172** 2894
[11] Hung C D. Johnson K and Torpy F 2006 *In Vitro Cell. Dev. Biol.* **42** 548
[12] Etienne H and Berthouly M 2002 *Plant Cell Tiss. Org. Cult.* **69** 215
[13] Hvoslef-Eide A K, Munster C, Heyerdahl P H, Lyngved R and Olsen O A S 2003 *Acta Hort.* **625** 173
[14] Deo P C, Tyagi A P, Taylor M, Harding R and Becker D 2010 *The South Pacific J. Nat. Appl. Sci.* **28** 27
[15] Rival A, Arbelenc-Bertossi F, Morcillo F, Tregear J, Verdeil J L and Duval Y 1997 *Plant Tiss. Cult. Biotechnol.* **3** 227
[16] Saenz L, Azpeitia A, Chuc-Armendariz B, Chan J L, Verdeil J L. Hocher V and Oropeza C 2006 *In Vitro Cell. Dev. Biol.* **42** 19
[17] Alkhateeb A A 2006 *Biotechnol.* **5** 466
[18] Steinmacher D A, Krohn N G, Dantas A C M, Stefenon V M, Clement C R and Guerra M P 2007 *Ann. Bot.* **100** 699
[19] Tahardi J S, Sianipar N F and Riyadi I 2002 *New Frontiers of Sago Palm Studies,* Universal Academic Press Inc., Tokyo p 75
[20] Jones M P A, Yi Z, Murch S J and Saxena P K 2007 *Plant Cell Rep.* **26** 13
[21] Murashige T and Skoog F 1962 *Physiol. Plant.* **15** 473
[22] Johansen D A 1940 *Plant Microtechnique* McGraw-Hill Book Company, New York
[23] O'Brien T P, Feder N and Mccully M E 1965 *Protoplasma* **59** 368
[24] Dam A, Paul S and Bandyopadhyay T K 2010 *Sci. Hort.* **126** 253

[25] Ramakrishna A, Giridhar P, Jobin M, Paulose C S and Ravishankar G A 2011 *Plant Cell Tiss. Organ Cult.* **108** 267
[26] Sumaryono and Riyadi I 2005 *Menara Perkebunan* **73** 1
[27] Riyadi I 2010 *J. Agro. Biogen* **6** 101
[28] Kasi P D and Sumaryono 2006 *Menara Perkebunan* **74** 44
[29] Hajong S, Kumaria S and Tandon P 2013 *J. Agric. Sci. Technol.* **15** 1425
[30] Karimpour S, Davarynejad G H, Bagheri A and Tehranifar A 2013 *J. Agric. Sci. Technol.* **15** 1209.
[31] Gaillochet C and Lohmann J U 2015 *Development* **142** 2237
[32] Alcantara G B D, Dibax R, Oliveira R A D, Filho J C B and Daros E 2014 *Act. Sci.* **36** 63
[33] El Dawayati M M, Ola H A E, Zaid Z E and El Din A F Z 2012 *Ann. Agric. Sci.* **57** 117

IOP Conf. Series: Earth and Environmental Science **183** (2018) 012008 doi:10.1088/1755-1315/183/1/012008

Somatic embryogenesis of sago palm (*Metroxylon sagu* Rottb.) from different origins in Indonesia

Masna Maya Sinta[1*], I Riyadi[1] and Sumaryono[1]

[1]Indonesian Research Institute for Biotechnology and Bioindustry, Jalan Taman Kencana No.1, Bogor 16128, Indonesia

*E-mail: mayasinta77@gmail.com

Abstract. Sago palm (*Metroxylon sagu* Rottb.) is considered as one of the prospective carbohydrate-producing crops in Indonesia. The development of commercial sago plantations has been hampered by the shortage of superior planting materials in a large quantity. Rapid and large-scale propagation of superior varieties could be accomplished by tissue culture. The Indonesian Research Institute for Biotechnology and Bioindustry has developed tissue culture of sago through somatic embryogenesis. The research objective was to determine the response of sago origins to previously developed somatic embryogenesis procedure. Explants of young suckers were taken from six different sago areas in Indonesia: Riau, West Java, South Kalimantan, Moluccas, Papua, and Southeast Sulawesi. Callus initiation was conducted on a solid modified-MS medium. The frequency of callus initiation was low and highly varied between 0.7% (Southeast Sulawesi and Papua) and 12.5% (West Java). The first callus was formed at the fourth culture (20 weeks) on explants from West Java and South Kalimantan, and at the sixth culture (29 weeks) on explants from Southeast Sulawesi. Callus multiplication rate was generally very high, 2-3.4 folds in 4 weeks. Calli produced from all explant origins could be successfully developed for somatic embryos and plantlets.

1. Introduction

Sago palm (*Metroxylon sagu* Rottb.) is one of the most potential carbohydrate-producing crops in the tropical area. Sago palm is monocarpic in which at 10-15 years of age produces a large branched terminal inflorescence above the leaves. The starch is stored in the trunk and harvested just before flowering. Sago palm yields 15-25 ton dry starch/ha/year [1]. Sago starch has long been a staple food for people in many parts of Indonesia such as Mentawai islands, Moluccas, and Papua. In addition to being a raw material for food, sago starch is also used in the production of white bread, noodles, high-fructose syrup, biodegradable filler in plastics, adhesive, bioethanol, animal feed, and many other derivative products [1].

The palm is distributed in tropical Southeast Asia mostly in Indonesia and Papua New Guinea, also in Malaysia, Thailand, Philippines, and the Melanesian islands of the Pacific Ocean. Sago in Indonesia is spread over several sago areas, such as in Java, Riau, Moluccas, Papua, Kalimantan, Sulawesi and Mentawai islands. Sago is able to grow in areas where most other crops cannot grow well such as in the riverbanks, lowland swampy areas, peat soils, flooded lands, and coastal lowland areas.

Sago palm can be propagated using suckers and seeds. Seed production is rare because the trees are harvested prior to the onset of flowering; therefore, suckers commonly propagate this sago palm.

However, to establish large-scale sago plantations, the availability of uniform suckers in a large number is a major constraint [2]. Tissue culture is a promising alternative means for rapid and mass clonal propagation of superior sago palms. In addition, tissue culture can be used as a method of *in vitro* conservation of sago palm. Tissue culture of sago palm through somatic embryogenesis has been successfully established by the Indonesian Research Institute for Biotechnology and Bioindustry (IRIBB) [3, 4]. Apical meristematic tissues of young suckers of selected sago palms were used as explants. The research objective was to determine the effect of plant origins on somatic embryogenesis of sago palm.

2. Materials and Methods

2.1. Explants and culture conditions

The research was conducted at the Laboratory of Plant Cell Culture and Micropropagation, IRIBB, Bogor from 2001 to 2010. Apical shoot tissues of young suckers of the field grown sago palms in six sago-producing areas: Bogor (West Java), Banjarmasin (South Kalimantan), Selat Panjang (Riau, Sumatra), West Seram (Moluccas), Merauke (Papua), and Kendari (Southeast Sulawesi) were used as explant materials (**Figure 2a**). All cultures except for callus initiation were incubated in the light culture room under white fluorescent lamps providing 20 μmol photon/m^2/s over a 12-h photoperiod at a constant temperature of 26 ^{o}C. Culture medium was sterilized in an autoclave at 121^{0}C for 20 min before being used.

2.2. Callus culture

Apical shoot tissues of young suckers were cultured on a modified MS (MMS) medium [3]. The cultures were placed in a dark culture room and were sub cultured every 6-8 weeks. Callus formed was sub cultured on the same medium with a lower concentration of 2,4-D for callus proliferation under light conditions.

The primary calli formed were sub cultured on solid MMS media in jar bottles added with 10-20 mg/L 2,4-D and 0.1 mg/L kinetin for callus proliferation. These cultures were placed in a dark room for 6 to 8 weeks. Calli formed were used as a material source for somatic embryo induction or for further callus proliferation.

2.3. Somatic embryo induction on a solid medium

The embryogenic callus was cultured on MMS solid media supplemented with 0.1 mg/L kinetin and 5 mg/L 2,4-D. The cultures were placed in a light culture room for 4 weeks when the frequency of somatic embryo formation was determined.

2.4. Somatic embryo maturation and germination in TIS

Somatic embryos formed were cultured on a liquid MMS medium with a half-strength of macro salts, 0.01 mg/L ABA, 1 mg/L kinetin and 0.1 mg/L GA_3. Flask of TIS consists of an upper containment for the somatic embryos and a lower containment for the media (**Figure 2d**). The volume for liquid culture was 75-100 ml per flask. To avoid contamination, 15 mg/l rifampicin and 15 mg/l tetracycline were added into the medium. When the lower containment was pressured, liquid medium was pumped into the upper containment immersing the plant culture for a specific period and interval of immersion adjusted automatically. The somatic embryos were immersed for 3 min every 6 h.

Somatic embryos at advanced developmental stages of scutellar and coleoptillar were subcultured in liquid MMS media supplemented with 0.5 mg/L GA_3 and 0.5-1.0 mg/L kinetin. The cultures were placed in a light culture room for 4-6 weeks. TIS was adjusted with 3 min immersion duration and immersion interval every 6 or 12 h [5].

IOP Conf. Series: Earth and Environmental Science **183** (2018) 012008 doi:10.1088/1755-1315/183/1/012008

2.5. Plantlet development

The resulting germinant were sub cultured on an MMS solid medium or in a liquid medium in the test tubes added with 0.5-1.0 mg/L GA_3 and 1.0 mg/L kinetin. The cultures were placed in a light room for 6 - 8 weeks [6]. Healthy vigorous plantlets with height more than 10 cm, had more than 3 leaves and good root systems were ready for acclimatization in *ex vitro* conditions.

3. Results and Discussion

3.1. Origin and description of sago palm

Sago palms collected from six regions in Indonesia had different local names (**Table 1, Figure 1a-f**). The sago palm trees had spines or without spines (spineless). The spines of sago palm were usually not more than 4 cm in length. This spininess characteristic is important for sago cultivation because sago with the spine is more difficult in managing the suckers than that of spineless one. In addition, explants collection from spined sago was more difficult than unspined sago palms.

In this research, unspined sago palms were collected from West Java, South Kalimantan and Papua, while spined sago palms were from Riau and Southeast Sulawesi. Both types of sago palms were obtained from West Seram, the Moluccas with different local names, Tuni for spined sago and Molat for unspined sago (**Table 1**).

Table 1. Origin and description of sago palm collected for tissue culture.

Explant Origin	District	Date of Collection	Local Name	Spines
West Java	Bogor	May 2001	Kirai	No
South Kalimantan	Banjarmasin	August 2004	Sagu	No
Riau	Selat Panjang	August 2007	Duri	Yes
Moluccas	West Seram	July 2008	Tuni	Yes
Moluccas	West Seram	July 2008	Molat	No
Papua	Merauke	November 2008	Alitir	No
Southeast Sulawesi	Kendari	October 2009	Runggamano	Yes

Figure 1. Sago palm from different origins: (a). West Java, (b). South Kalimantan, (c). Riau, (d). Moluccas, (e). Papua, and (f). Southeast Sulawesi.

3.2. Callus culture

Explants from different origins responded differently to the media used. The first callus was formed at the fourth culture (20 weeks), the fastest response on explants from West Java and South Kalimantan, and at the sixth culture (29 weeks) the slowest response on explants from Southeast Sulawesi (**Table 2**). The frequency of callus initiation was low ranging from 0.7% (Papua and Southeast Sulawesi) to 12.5% (West Java). White friable callus emerged from the edge of the explants (**Figure 2b**). There was no correlation between spined or unspined sago on callus formation.

The results show that callus formation on explants from eastern region (West Java, South Kalimantan and Riau) was slower and lower in frequency than those of explants from the western region (Moluccas, Papua and Southeast Sulawesi). All calli formed were friable and white in color (**Figure 2b**) with proliferation rate between 2 to 3.4 folds (**Table 2**). Sago from Riau had relatively low callus formation (2%) but the calli proliferation rate was high.

Table 2. Callus culture of different sago origins.

Explant origin	First callus emerge (week)	Callus formation (%)	Callus type	Nodular callus formation (%)	Nodular callus proliferation (times)
West Java	20	12.5	Friable, white	100	3.0
South Kalimantan	20	1.4	Friable, white	100	3.0
Riau	24	2.0	Friable, white	100	3.4
Moluccas Tuni	27	1.4	Friable, white	100	3.0
Moluccas Molat	27	0.9	Friable, white	100	2.0
Papua	26	0.7	Friable, white	100	2.0
Southeast Sulawesi	29	0.7	Friable, white	100	2.0

3.3. Callus proliferation

Calli began to multiply at 4 weeks after culture. Callus biomass increased 3-5 times of the initial biomass. The biomass increase in TIS culture was much higher than that of on solid media culture. In TIS cultures, biomass increase at 6 weeks of age (at harvest time) reached up to 6.5 times of the initial biomass [5]. The calli from this proliferation stage were used as material sources for somatic embryo induction culture both on solid media and in TIS.

The rate of callus proliferation varied with sago origins. In general, the proliferation rate of sago callus originating from the western region was faster and higher than that of from the eastern region. The highest proliferation rate was obtained from sago plants originated from Riau at 3.4 times in 6 weeks (**Table 2**). The low frequency of callus initiation of sago is not a limitation in sago propagation using somatic embryogenesis technique. This condition can be reinstated with high sago callus proliferation rate, and the ability of calli to form somatic embryos that reached 100%.

3.4. Somatic embryo induction and regeneration

All nodular calli of sago can be induced into somatic embryos (**Figure 1c**). The difference between the sago types and origins was on emergent time and proliferation rate of somatic embryos. Sago origin of Riau and Kalimantan showed the fastest and highest response on induction of somatic embryos. On the other hand, sago origin of Southeast Sulawesi and Papua showed the slowest response on somatic embryo induction and proliferation.

Maturation and germination of somatic embryos of sago were successfully conducted both on solid and in TIS media (**Figure 2**). TIS allows explants to be exposed to the culture media periodically in a very short time with a good air circulation. The efficiency of maturation and germination of somatic embryos was almost the same for all sago origins from both western and eastern Indonesia. TIS has been used successfully for SE of oil palm [7]. In case of flowering abnormality such as mantled fruits in oil palm due to the long-term duration of the culture, is not a problem in sago plants because sago palms are harvested before flowering.

Secondary somatic embryos (SSE) appeared during somatic embryo maturation stage [4]. The formation of SSE has been occurred subsequently at the germination stage of somatic embryos, especially on sago originated from Riau, Kalimantan, Moluccas (Tuni), and Papua (Alitir) [6]. Plantlet development and rooting were conducted on solid or in liquid media before acclimatization. Selection of plantlets to be acclimatised was done by determining the criteria of plantlets that had more than 3 leaves, the height was more than 7 cm, and had a good root system.

Plantlets of sago were successfully acclimatized in the nursery. The survival rate of plantlets on average was 40-60% in all sago origins. For the commercial purposes, this survival rate frequency is considered low, therefore more efforts should be done to increase sago plantlets survival rate. The most critical period in the acclimatization stage was in the first stage of acclimatization where the plantlets were placed in a closed plastic tunnel for 12 weeks. When this period was successfully accomplished, the plantlets had a high chance to survive afterwards. In the second stage, the plantlets were transferred to larger polybags (30 cm x 25 cm) and placed below the 50-60% shading plastic net for 6-8 weeks and then without shading net for 6-8 weeks. After this stage, sago plants were ready to be transplanted to the field.

Figure 2. Sago palm tissue culture: (a). Young suckers as explant sources, (b). Callus emerged from shoot apical explants, (c). Somatic embryos, (d). Embryo maturation on TIS, (e). Plantlets, and (f). Tissue culture derived sago plants in the nursery.

4. Conclusions

Callus initiation of sago was varied according to the plant origins. Sago explants from the western region of Indonesia formed callus earlier and in higher frequency than those of eastern regions. Sago explants from West Java had the best response for callus initiation. The proliferation rate of nodular callus was different depending on explant origins where explants from Riau was the fastest in callus proliferation. Calli from all explant origins could be converted to somatic embryos and plantlets.

5. Acknowledgments

This research was a part of the Applied Research Incentive Program funded by the Ministry of Higher Education, Research and Technology, Republic of Indonesia.

IOP Conf. Series: Earth and Environmental Science **183** (2018) 012008 doi:10.1088/1755-1315/183/1/012008

6. References

[1] Flach M 1997 Sago Palm *Metroxylon sagu* Rottb. Promoting the conservation and use of underutilized and neglected crops 13. *International Plant Genetic Resources Institute*. Rome, Italy. 76pp
[2] Jong F S 1995 Research for the development of sago palm (*Metroxylon sagu* Rottb.) cultivation in Sarawak, Malaysia *Sadong Press Sdn*. Bhd. 139pp
[3] Tahardi J S, N F Sianipar and I Riyadi 2002 Somatic embryogenesis in sago palm (*Metroxylon sagu* Rottb.), p.75-81 In K Kaimuna, M Okazaki, Y Toyoda and J E Cecil (Eds.). *New Frontiers of Sago Palm Studies*. Universal Academy Press, Tokyo, Japan
[4] Riyadi I, J S Tahardi and Sumaryono 2005 *Menara Perkebunan* **69** 46
[5] Kasi P D and Sumaryono 2008 *Menara Perkebunan* **76** 1
[6] Riyadi I, Efendi D, Purwoko B S and Santoso D 2016 Somatic embryogenesis of sago palm (*Metroxylon sagu* Rottb) using liquid culture method for technological development of elite clone seedling propagation [*Dissertation*] Bogor: Agricultural University 97pp
[7] Sumaryono, Riyadi I, Kasi P D and Ginting G 2008 *Indon. J. Agric.* **1** 109

IOP Conf. Series: Earth and Environmental Science **183** (2018) 012009 doi:10.1088/1755-1315/183/1/012009

Embryogenic callus initiation from leaf explants of *Elaeis oleifera* x *Elaeis guineensis* (OxG) hybrids

Sumaryono[1]*, I Riyadi[1], R T Saptari[1], H Y Rahmadi[2] and Ernayunita[2]

[1]Indonesian Research Institute for Biotechnology and Bioindustry, Jalan Taman Kencana No.1, Bogor 16128, Indonesia
[2]Indonesian Oil Palm Research Institute, Jalan Brigjen Katamso No.51, Medan 20158, Indonesia

*E-mail: sumaryono@iribb.org

Abstract. *Elaeis oleifera* x *Elaeis guineensis* (OxG) hybrid is an oil palm interspecific crossbreed to produce a variety with better oil quality, slow stem growth, and resistance to bud rot disease. However, one of the problems in OxG hybrid breeding program is an inefficiency in the conventional propagation method, while the protocol for OxG hybrid tissue culture has not been established yet in Indonesia. Somatic embryogenesis (SE) using a temporary immersion system (TIS) will be used for OxG clonal propagation as it has been established successfully in *E. guineensis*. Explants were young leaves derived from genotypes of selected F1 and backcross (BC) of OxG hybrids where the *E. oleifera* was originated in Brazil and Suriname. Callus initiation used eight media with various compositions of minerals, hormones, and active charcoal. Initial callus emerged in 8-10 weeks after culture. The average of callogenesis frequency was 3.49% in F1 and 5.59% in BC hybrid explants, with the highest at 10.73% occurred in BC Brazil. Callogenesis of OxG hybrid showed a high potency as callogenesis of *E. guineensis*. The callus had nodular structure which was potentially embryogenic. After 3 cultures, some of the embryogenic calli were transferred to TIS for somatic embryo induction.

1. Introduction

Oil palm belongs to the Arecaceae family and is the largest and the most efficient source of vegetable oil in the world. Palm oil is mainly used as an edible product, but also has many other uses such as non-food derivatives, oleochemicals, and biofuels. The demand for palm oil has increased considerably especially due to the increasing demand for biofuels and for edible oils especially in Asia [1]. In order to meet this demand, the oil palm plantation areas in many oil palm producing countries, including Indonesia are expanding rapidly.

There are two commercial oil palms, African oil palm (*Elaeis guineensis* Jacq.) and American oil palm (*Elaeis oleifera* HBK Cortes). Due to its higher yield, African oil palm, *E. guineensis,* is overwhelmingly used for commercial cultivation. The fast expansion of *E. guineensis* as monoculture plantations in very large areas is dangerously fragile to climate changes and diseases attacks; therefore crossbreed with *E. oleifera* that has useful genetic traits is very important in oil palm breeding program. *E. oleifera* has a low oil yield, but it has valuable characters such as a low shoot growth rate, a high content of unsaturated fatty acids, substantial levels of carotenes and vitamin E, and more resistance to bud rot disease. An interspecific crossbreed of *E. oleifera* and *E. guineensis* (OxG) is intended to combine both valuable traits of the parents and its backcross (BC) with *E. guineensis* is expected to maintain higher yields.

OxG hybrids have been developed mostly in the Latin American countries such as Ecuador, Colombia and Brazil [2] to overcome yellowing disease which *E. guineensis* is susceptible while *E. oleifera* is resistance but with lower oil yield. OxG hybrids have also been developed in Southeast Asian countries for slower growth and better oil quality. The physiological and morphological characteristics of *E. oleifera* and their hybrids (OxG) have been determined to select the best female parents of OxG hybrids for breeding program purposes and the best hybrids for agronomic performance [3]. The best progenies of OxG hybrids and BC should be clonal propagated to maintain their characteristics. These OxG hybrids have a problem regarding the development of embryos, resulting occasionally in seed abortion. Thus, propagation of OxG hybrids by seeds is limited. Since oil palm belongs to monocotyledonous plant and does not produce any suckers, its clonal propagation can be conducted only by tissue culture.

In vitro propagation of OxG hybrids through somatic embryogenesis (SE) has been reported mostly using zygotic embryos as explants [4, 5, 6], but SE in OxG hybrids from young leaf explants has been reported only by Ginting *et al.* [7] using one OxG hybrid on solid media. However, SE with young leaves as explants has been widely used in *E. guineensis* [8]. Furthermore, SE using a temporary immersion system (TIS) was reported to improve somatic embryo production and uniformity [9] and decrease floral abnormality in *E. guineensis* [10]. Therefore, in this research SE-TIS was applied for clonal propagation of F1 and BC of OxG hybrids developed by the Indonesian Oil Palm Research Institute (IOPRI) where the *E. oleifera* was originated from Brazil and Suriname. At the present time, we report the progress of the callus initiation and embryogenic calli formation from young leaf explants of OxG hybrids.

2. Materials and Methods

2.1. Plant material and culture conditions

The research was conducted at the Laboratory of Plant Cell Culture and Micropropagation, the Indonesian Research Institute for Biotechnology and Bioindustry (IRIBB), Bogor, West Java. Explants used were young leaves of selected genotypes of F1 and BC of OxG hybrids that have been developed by IOPRI since the 1970's where the *E. oleifera* plants were originated from Brazil and Suriname. The OxG genotypes were planted at Marihat and Bah Jambi experimental gardens, Pemantang Siantar, North Sumatra, Indonesia. The explant sources were selected from the best progenies of F1 and BC OxG hybrids of Brazil and Suriname origins. Selection of the progenies was based on fruit productivity, oil quality and vegetative growth. The explants were cut from mature trees, cleaned, washed with alcohol, packed, and sent to Bogor by air. Young leaves used were from leaf phyllotaxis number from -3 to -8. Approximately 3000 explant cuts could be generated from one young leaf explant of oil palm.

All the cultures, otherwise stated, were placed in a dark culture room at a constant temperature of 26±2 °C.

2.2. Treatments and observation

Every leaf explant cut (1 cm x 2 cm) having a mid-vein with two blades was placed on a small culture tube (diameter 2.5 cm, height 15 cm) containing a solid medium with various compositions of minerals, hormones, and active charcoal. Eight different media (M1 to M8) were used as treatments for callus initiation of selected genotypes of F1 and BC of OxG hybrids from Brazil and Suriname (F1 Brazil, F1 Suriname, BC Brazil, BC Suriname). Each genotype consisted of at least 90 explant cuts (tubes) of each medium treatment. The pH of the medium was adjusted to 5.8 before autoclaved at 121 °C and 1.0 kg/cm^2 for 20 min.

The observation was conducted every 2 weeks in the first two months on contamination and callus emergence, and then every month until 8 months on callus formation. The cultures without callus

IOP Conf. Series: Earth and Environmental Science **183** (2018) 012009 doi:10.1088/1755-1315/183/1/012009

formation were discarded after 8 months of culture. Primary calli that have been formed were transferred to solid media with a reduced concentration of auxin for callus proliferation and differentiation. After 3 subcultures, the embryogenic calli were transferred to liquid culture in a temporary immersion system (TIS) using 250 mL Nalgene flasks for somatic embryo induction. The embryogenic calli were immersed for 3 min every 6 h [10]. The cultures were incubated in the culture room at 10 $\mu mol/m^2/s$ light intensity over a 12-h photoperiod.

3. Results and Discussion

3.1. Callus initiation

Leaf explants folded along the vein and became green brownish. Initial calli emerged in 8 weeks, quicker than those of *E. guineensis* where usually emerged in 10-12 weeks [8, 11]. The primary callus together with its leaf explant was sub cultured to a new media after 4, 6, and 8 months of culture. The callus was white brownish in colour and its texture was mostly soft and watery (**Figure 1**).

The average of callogenesis was 3.49% in F1 and 5.59% in BC hybrid explants on eight different media, after 8 months of culture. BC Brazil explants had the highest callogenesis frequency on average at 10.73%, followed by F1 Suriname at 5.41%, F1 Brazil at 1.73% and BC Suriname at 0.39% (**Table 1**). Hence, the best media for callogenesis of OxG hybrids were M6 and M3 at 8.87% and 8.11%, respectively. Leaf explants of F1 Brazil and BC Suriname cultured on several media did not form any calli at all. On the contrary, leaf explants of BC Brazil on M6 and M3 produced callus at the frequency of 18.89% and 18.21%, respectively. M6 media also gave the high frequency of callus initiation on F1 Suriname at 16.23%. This callogenesis frequency of OxG hybrid was comparable to that of *E. guineensis* at 15% [8] or 11-20% [12] and another OxG hybrid at 11.25% [7]. Callogenesis frequency of OxG hybrids using zygotic embryos was very high at 77.9-95.3% after 90 days on Brazilian OxG hybrids [4] and at 10-60% after 6 months on other Brazilian OxG hybrids [6]. Activated charcoal at 0.25% was used to reduce oxidation in the culture of zygotic embryos of OxG hybrids [13].

Callogenesis levels can be improved by selecting oil palms as a source of explants (ortets) that are predicted by molecular markers to be amenable to tissue culture process. Gene expression markers for callogenesis of oil palm have been identified to be associated with cytokinin and brassinosteroid such as putative cytokinin dehydrogenase (*EgCKX*), brassinosteroid responsive gene (*EgBrRK*), and putative response regulator type A gene (*EgRR1*) [14]. These markers were significantly correlated to callogenesis rates of oil palm, therefore can be used as an early assessment of the callogenesis potential of oil palms [14].

Table 1. Callogenesis frequency of OxG hybrids on different culture media (%).

Culture media	OxG Hybrids				Average
	F1 Brazil	F1 Suriname	BC Brazil	BC Suriname	
M1	3.94	7.34	14.68	0.34	6.58
M2	4.52	1.59	7.25	0.00	3.34
M3	3.04	9.17	18.21	2.00	8.11
M4	0.00	1.56	6.06	0.00	1.90
M5	0.68	0.96	1.52	0.00	0.79
M6	0.35	16.23	18.89	0.00	8.87
M7	0.00	1.05	6.96	0.34	2.09
M8	0.41	4.92	12.81	0.33	4.62
Average	1.73	5.41	10.73	0.39	4.54

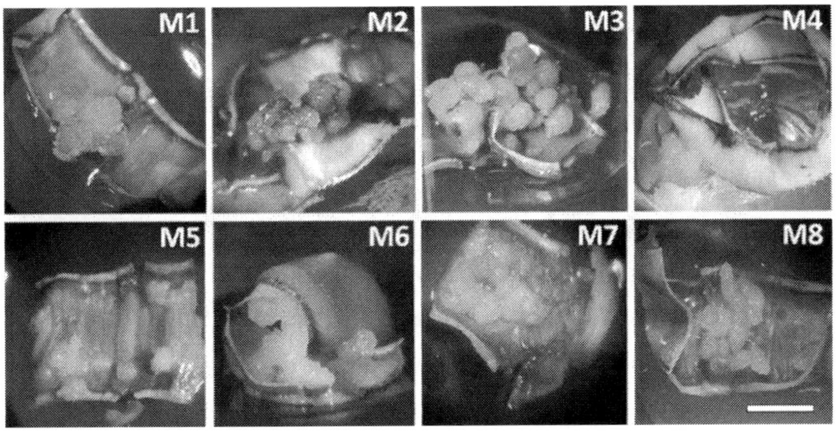

Figure 1. Callus formation in BC OxG hybrid Brazil on different media after 4 months. Bar = 0.5 cm.

3.2. Primary callus culture

The primary calli formed after 4 months, 6 months and 8 months were transferred to solid media containing a lower concentration of auxin. After 3 cultures, there was no significant increase in the callus biomass in term of the number of callus clumps, but the callus was growing bigger. Some calli were contaminated or dead, reducing the number of callus clumps. After 8 months, OxG hybrid BC Brazil had 308 callus clumps (culture tubes), F1 Suriname had 111 clumps, F1 Brazil had 37 clumps, and BC Suriname had only 7 callus clumps (**Table 2**). Many of the calli become more friable (**Figure 2A**) and some calli turned to embryogenic and nodular in structure (**Figure 2B**) after being subcultured several times.

The calli with nodular structure (**Figure 2B**) were very potential to form somatic embryos. These calli also grew relatively fast. Ginting *et al.* [7] restricted to use only calli formed within the first 5 months of OxG hybrid culture. Primary callus culture was conducted for 4 months to form somatic embryos. However, only a limited number of somatic embryos were obtained [7].

Table 2. Number of callus clumps of OxG hybrids on different culture media after 8 months.

Culture media	OxG Hybrids				Total
	F1 Brazil	F1 Suriname	BC Brazil	BC Suriname	
M1	12	17	54	0	83
M2	13	5	26	0	44
M3	8	26	63	5	102
M4	0	4	25	0	29
M5	2	2	3	0	7
M6	1	47	71	0	119
M7	0	2	23	1	26
M8	1	8	43	1	53
Total	37	111	308	7	463

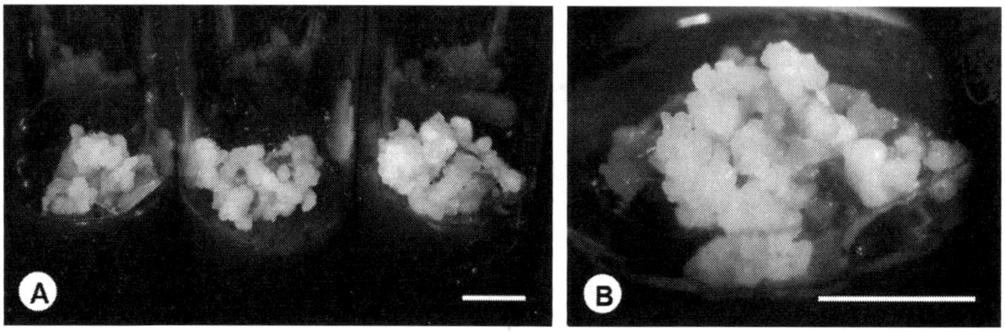

Figure 2. Embryogenic calli of OxG hybrid after the second (A) and the third (B) primary callus culture. Bar = 1 cm.

3.3. Somatic embryo induction in TIS

After 4 cultures, the embryogenic calli were transferred to TIS. Somatic embryo induction, maturation, and germination will be conducted in TIS. Liquid culture using TIS has been utilised successfully in *E. guineensis* [9, 10] but has not been applied either in *E. oleifera* or OxG hybrids. In *E. guineensis,* SE using TIS was stated to improve somatic embryo production and uniformity [9] and to decrease floral abnormality [10]. The embryogenic calli were in TIS for the first passage and have not been converted to somatic embryos yet.

Figure 3. Induction of somatic embryos in TIS. Bar = 1 cm.

4. Conclusions

Calli have been successfully initiated from F1 and BC of OxG hybrids of oil palm. Different OxG hybrids responded differently to different media of callus initiation. The best callogenesis was BC OxG hybrid Brazil at 18.9% on M6 medium and at 18.2% on M3 medium.

After being sub cultured on solid media with a lower concentration of auxin, the primary callus has become embryogenic and nodular in structure. These embryogenic calli have been transferred into a liquid culture of TIS. The embryogenic calli on TIS have not produced somatic embryos yet.

5. Acknowledgements

This research was a part of Project No.PRJ-51/DPKS/2016 funded by the Oil Palm Estate Crop Management Fund Agency (BPDP Kelapa Sawit), Ministry of Finance, Republic of Indonesia.

6. References

[1] Murphy D J 2014 *J. Oil Palm Res.* **26** 1

[2] Kushairi A, Tarmizi A H, Zamzuri I, Ong-Abdullah M, Samsul K R, Ooi S E and Rajanaidu N 2010 *International Seminar on Advances in Oil Palm Tissue Culture*, Yogyakarta, Indonesia p 1

[3] Rivera Y D. Cayón D G and López J E 2013 *Agro. Colombiana* **31** 314

[4] Silva Angelo P C, Lopes R, Moraes L A C and da Cunha R N V 2009 *Crop Breed. Appl. Biotechnol.* **9** 274

[5] Silva Angelo P C, Steinmacher D A, Lopes R, da Cunha R N V and Guerra M P 2013 *Agric. Sci.* **4** 1

[6] Bonetti K A P, Nesi J, Quisen R C and Quoirin M 2016 *African J. Biotechnol.* **15** 2028

[7] Ginting G, Fatmawati and Purba A R 1994 *Buletin PPKS* **2** 17

[8] Soh A C, Wong G, Tan C C, Chew P S, Chong S P, Ho Y W, Wong C K, Choo C N, Nor Azura H and Kumar K 2011 *J. Oil Palm Res.* **23** 935

[9] Tahardi J S 1998 Proc. *Internat. Oil Palm Conf.* Bali: Indonesia p 595

[10] Sumaryono, Riyadi I, Kasi P D and Ginting G 2008 *Indonesian J. Agric.* **1** 109

[11] Yusnita and Hapsoro D 2011 *Hayati J. Biosci.* **18** 61

[12] Ho Y W, Tan C C, Soh A C, Wong G, Chong S P, Choo C N and Norazura A 2009 *International J. Oil Palm* **6** 86

[13] Alves S A O, de Lemos O F, dos Santos Filho B G and da Silva A L L 2011 *J. Biotechnol. Biodivers.* **2** 1

[14] Ooi S E, Choo C N, Zamzuri I and Ong-Abdullah M 2013 *J. Oil Palm Res.* **25** 9

Optimization of simultaneous saccharification and fermentation in bioethanol production from sugarcane bagasse hydrolyse by *Saccharomyces cerevisiae* BTCC 3 using response surface methodology

Ahmad Thontowi[1*], U Perwitasari[1], L N Kholida[1], Fahrurrozi[1], Yopi[1] and B Prasetya[2]

[1]Research Centre for Biotechnology, Indonesian Institute of Sciences (LIPI)
[2]National Standardization Agency, Indonesia

*E-mail: ahmad.thontowi@lipi.go.id

Abstract. The response surface method (RSM) was used to investigate the effects of bioethanol fermentation parameters, including sugarcane bagasse concentrations (0-5 % (w/v), enzyme types (cellulase and hemicellulase), enzyme concentrations (0-1 FPU/g biomass), medium pH, and incubation period. Bioethanol production was conducted by simultaneous saccharification and fermentation (SSF) of sugarcane bagasse hydrolyse using *Saccharomyces cerevisiae* BTCC 3. The experiment results showed that the maximum bioethanol production by using the SSF method was the same as RSM prediction. The optimum conditions for bioethanol production were 5% (w/v) of sugarcane bagasse, a mixture of cellulase and hemicellulase at 1 FPU/g of sugarcane bagasse for each enzyme, pH medium 6.0, and fermentation period 72 hours. The bioethanol yield of 2.43 g/L was obtained under these conditions.

1. Introduction

Bioethanol is one of the biofuels that mostly used in the transportation sector. The USA and Brazil are the largest producers of bioethanol in the world with 52% and 37% production, respectively [1]. At present, bioethanol for the fuel market is produced from corn in the United States and from sugarcane in Brazil. These food grade raw materials will not be sufficient to meet the increasing demand for bioethanol fuel. In addition, it accommodates greenhouse gases using bioethanol based on flour or sugar is not as high as desired [2]. On the other hand, bioethanol can also be produced from abundant and renewable biomass sources such as organic wastes of agricultural crops, forests, and other crops [3, 4, 5].

For the large-scale biological production of bioethanol, it is desirable to use cheap and abundant substrates. Bioethanol production cost using corn (the main component of starch) or sugarcane (the ingredients in cane juice or molasses) is 40-70% higher than that of raw materials [6, 7]. Bioethanol production cost is cheaper by using lignocellulose waste products from the forestry, agriculture, and industry [5]. Lignocellulose is a complex polymer consisting of three carbohydrates, namely hemicellulose, cellulose and lignin. This polymer is considered an attractive feedstock for the production of fuel bioethanol since it is available in a large quantity at low cost [8, 9, 10]. In addition, it can reduce competition with food ingredients.

Several reviews have been published with the theme of bioethanol production from lignocellulosic biomass [8, 11, 12]. Lignocellulosic materials from residues of several different plants have been used and converted to bioethanol. Sugarcane bagasse is one of the major lignocellulosic biomass found in large quantities in the tropical countries. The fibrous residue was obtained after extracting the juice from sugarcane (*Saccharum officinarum*) in the sugar production process [13]. Indonesia is one of the producers of sugarcane bagasse [14, 15] as a raw material for bioethanol production.

The bagasse has a strong crystal structure formed of cellulose, hemicellulose, and lignin. The lignocellulosic sugarcane can be hydrolysed into sugars and fermented into bioethanol [16]. During simultaneous saccharification and fermentation (SSF), the sugars released from biomass are immediately broken down into bioethanol by yeast cells to reduce lead inhibition of enzymes and reduce of production costs [17]. During the SSF process, the ethanol content, concentration, and level of the fermentation products are strongly influenced by operating parameters, including solid biomass concentration, enzyme, temperature, and yeast concentration. Surface Response Methodology (RSM) could be used for analysis of bioethanol production using SSF from sugarcane bagasse including of the biometric series process, concentration and production rate. RSM includes three main methods: (1) implementation of a statistically experimental plan to collect data; (2) development of regression model to correlate experiment data; and (3) prediction of the target response variable against the process parameters using a regression model [18].

The purpose of this study was to analyse the regression model to predict the rate of bioethanol production, concentration and production rate during SSF sugarcane bagasse and to use the regression model to optimize the parameters of SSF process operation including the concentration of solid biomass, and the type and concentration of enzymes.

2. Materials and Methods

2.1. Microorganism and medium

Saccharomyces cerevisiae BTCC 3 was obtained from Biotechnology Culture Collection (BTCC), Research Centre for Biotechnology, Indonesian Institute of Sciences (LIPI). BTCC 3 strain was cultured in 40 mL of Yeast Peptone Glucose (YPD) medium at 30 °C and 150 rpm for 18 hours. The cells were centrifuged for 15 min for 10,000 g at 4 °C and then washed with aquadest. The cells were collected and diluted with 40 mL of aquadest then used as inoculum for the next study.

2.2. Pre-treatment of sugarcane bagasse biomass

Sugarcane bagasse hydrolysate was used as the substrate for bioethanol fermentation. The dried sugarcane bagasse was ground to 80 mesh of size, hydrolysed with 0.5M H_2SO_4, and then filtered to separate the filtrate from the biomass. Sugarcane bagasse hydrolysate with 5% moisture content was used as a substrate in this study.

2.3. Fermentation of BTCC 3 strain for ethanol production

SSF of sugarcane bagasse hydrolysate was carried out on a 100 mL Erlenmeyer flask with 12 mL fermentation medium. Fermentation was performed on a shaker incubator at 30 °C and 150 rpm. Yeast Nitrogen Base (YNB) without amino acid as much as 6.7 g/medium was used in this fermentation process. Cellulase and hemicellulase enzymes (Sigma-Aldrich) were used for the saccharification process of sugarcane bagasse.

2.4. Response surface method

Response Surface Method (RSM) was used to determine the optimum conditions of SSF of sugarcane bagasse. The experimental design was provided by Design-Expert 10®. The major composite designs used for bioethanol production by BTCC 3 yeast strain was carried out with five variables, namely concentration of sugarcane bagasse, enzyme type, enzyme concentration, pH medium, and incubation time for the SSF process (**Table 1**). Total experiments on the main composite design used 31 Erlenmeyer flasks.

Table 1. The main composite design for the SSF process for ethanol production by using BTCC3 yeast strain.

Variable	Code	Degree							
Bagasse hydrolysate (%)	A	0	1	3	5				
Type of enzyme	B	Control (0)	Cellulase (1)	Hemicellulase (2)	Cellulase-Hemicellulase (3)				
Enzyme concentration (FPU/g biomass)	C	0.2	0.4	0.6	0.8	1			
pH medium	D	4	4.8	5.5	6				
Time of incubation (hour)	E	0	6	12	24	30	36	48	72

3. Results and Discussion

Increasing bioethanol production was along with enhancing the biomass concentration, the enzyme concentration, and incubation period. The lignocellulosic biomass consists of cellulose, hemicellulose, and lignin [19]. Sugarcane bagasse could be converted into simple sugar by lignocellulosic enzymes. Furthermore, the sugar was utilized by BTCC 3 yeast strain to produce bioethanol. Based on the type of enzyme used, the highest ethanol yield was produced during fermentation using a mixture of cellulase and hemicellulase. The efficiency of lignocellulose biomass hydrolysis requires a mixture of enzymes due to their complex structure [20].

3.1. Ethanol production

The RSM analysis resulted in the production of bioethanol in the fermentation of sugarcane bagasse with the two-factor interaction (2FI) model equation. This model is used to predict bioethanol production as a function with five variables, namely A: the composition of sugarcane bagasse (%); B: type of enzyme; C: enzyme concentration; D: pH of the medium; and E: incubation time (hour). The equation for predicting bioethanol production was given by the RSM application as follows:

$$\text{Ethanol (g/L)} = 0.13 + 0.14A + 0.12B + 0.098C + 0.15D + 0.13E + 0.17AB + 0.15AC + 0.16AD + 0.15AE + 0.12BC + 0.15BD + 0.10\,BE + 0.16CD + 0.11\,CE + 0.16\,DE$$

Experimental data and ethanol production prediction were given in **Table 2**. The analysis of variance (ANOVA) shows that the value of squared regression (R^2) is 0.9282. This value means that 92.82% of bioethanol production by SSF with sugarcane bagasse hydrolysate could be explained by this model. This is evidenced by the parameters of the optimum conditions for bioethanol production from both of the actual experiments and the predictions were similar (**Table 2**), with the result of the experimental ethanol concentration (2.43 g/L) was slightly higher than the predicted value (2.21 g/L).

The result of the analysis using ANOVA shows that the p-value for this experimental model was <0.0001. This value indicates that the model was significant. The F value also shows an established model. If F value exceeds the tabulated value F at the 0.0001 probability level, the null hypothesis (H0) is rejected at the 0.0001 significance level. ANOVA analysis results show a significant level of five parameters and their interaction in ethanol production.

Table 2. Optimum conditions for ethanol production with SSF by BTCC3 yeast strain.

A Biomass concentration (%)	B Type of enzyme	C Concentration of enzyme (FPU)	D pH	E Time of incubation (hour)	Ethanol (g/L)		Error (%)
					Prediction	Actual	
5	3	1	6	72	2.21	2.43 ± 6.655E-005	6.655E-005

The fermentation of sugarcane bagasse to produce bioethanol could take place if the biomass can be converted into sugar, such as glucose, xylose, cellobiose, and others. Cellulase enzymes were used for this process. This study used two types of enzymes, namely hemicellulase and cellulase. The ratio of these two enzymes was 1:1. Based on the type of enzyme applied, the highest bioethanol concentration produces during fermentation using a cocktail of the enzymes was around 0.8 g/L. While the fermentation used a single type of enzyme, 0.16-0.46 g/L of bioethanol were produced by BTCC 3 yeast strain (**Figure 1A**). The optimal enzyme concentration was known at 1 FPU/g of biomass for each type with 0.4 g/L of bioethanol was produced (**Figure 1B**). Sugarcane bagasse hydrolysate was hydrolysed around 5% at an optimum concentration of enzymes and 0.4 g/L bioethanol was produced in this condition. At lower concentrations of sugarcane bagasse and the enzyme concentration has to produce bioethanol with a smaller content of 0-0.18 g/L. In addition, with optimum concentrations of enzymes was obtained bioethanol of 0.8 g/L in pH 6.0 of the medium (**Figure 1D** and **1E**). The optimum time of sugarcane bagasse required 72 hours of fermentation using BTCC 3 yeast strain (**Figure 1F**).

Bioethanol yield in this study was less than another study that could produce 38 g/L of bioethanol [21]. Production of bioethanol could be increased if the concentration of sugarcane bagasse, enzyme concentration, and incubation time were increased. Wang *et al.* fermented 7% (w/v) sugarcane bagasse and used 25 FPU/g-glucan [21]. The similar result was reported by Dasgupta *et al.* using *Kluyveromyces* sp. IIPE453 [22]. This strain produced bioethanol at 17.44 g/L in 7% (w/v) of sugarcane bagasse with additional cellulase enzyme. Other studies reported that the saccharification of sugarcane bagasse with a combination of cellulase and hemicellulase enzymes provided a better response to bioethanol production compared with using only one type of enzyme. The reason is sugarcane bagasse has cellulose, hemicellulose and lignin fibres that required several enzymes to hydrolyse the complex structures of lignocellulose [23, 24].

One other important factor in ethanol fermentation is the pH of the medium [25]. The results of this study indicate that the pH of the medium closer to neutral conditions would provide a positive response to ethanol production. Table 2 shows the ANOVA results for bioethanol production with a fairly high F value and a low Prob>F value<0.005. The value assumed that the predicted function equation for bioethanol production could be used [26]. Sugarcane bagasse fermentation to bioethanol production has an optimum pH medium around 6.0. This condition is related to the BTCC 3 yeast strain characters. BTCC 3 yeast has a pH optimum range from 2.5 to 7.0 (unpublished data). *S. cerevisiae* IFST-072011 also has an optimum pH of 5.0-6.0 in bioethanol fermentation [27]. The pH medium also affects the formation of proteins/enzymes in the cells. The enzyme plays an important role in the process of cell metabolism. In this study hemicellulase and cellulase from Sigma Aldrich have a pH optimum of 5.5-6.0. These results indicate that the suitability of optimum pH for yeast strain and optimum conditions of pH becomes important for the fermentation of sugarcane bagasse.

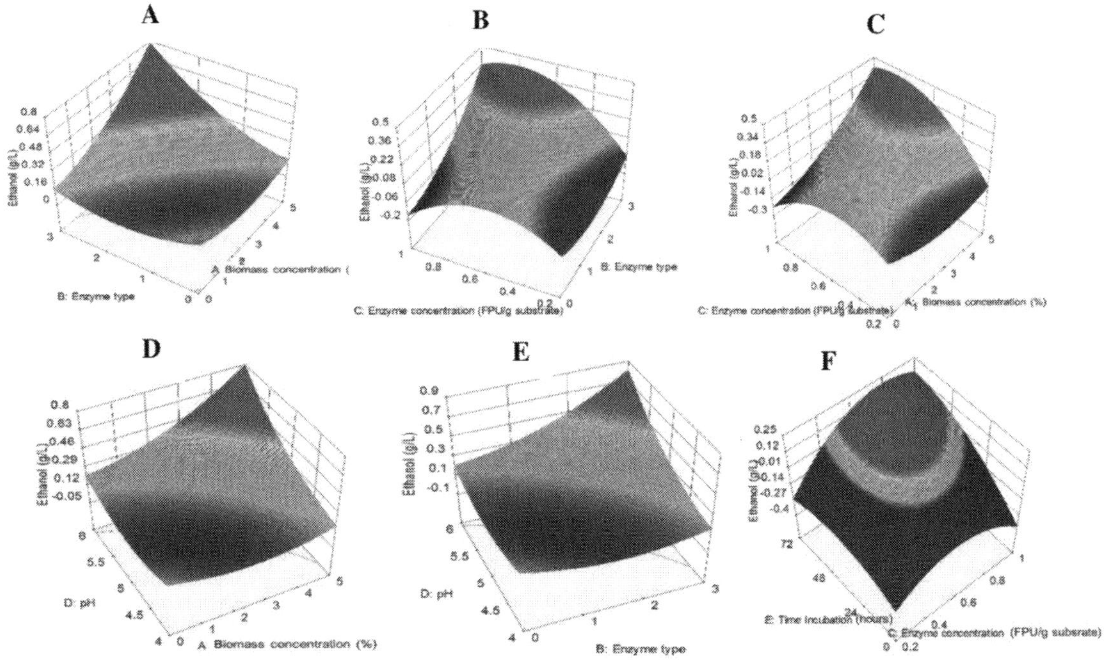

Figure 1. The response of ethanol production in several enzyme types (A) and enzyme concentrations (B), bagasse concentrations (C), media pH (D and E), and fermentation time (F). Fermentation took place at a temperature of 30 ° C and agitation speed of 150 rpm.

3.2. Glucose

Based on the ANOVA analysis, the p-value for the experimental model for glucose concentration as a fermentation product was 0.2329. This value indicated the model was insignificant. Significant levels of five parameters and interactions between parameters in glucose concentration were not significant. The application results provide a quadratic model for the glucose concentration as follows:

$$\text{Glucose (g/L)} = 19.06 + 0.29A - 2.65B + 3.01D + 1.03E + 0.67AB + 0.90AC - 4.66AD - 2.4AE + 1.15BC + 2.32BD - 0.15BE + 4.06CD - 0.77CE - 0.80DE - 11.70A2 + 10.51B2 - 0.12C2 + 0.79D2 - 3.84E2$$

Description: A: composition of sugarcane bagasse (%); B: type of enzyme; C: enzyme concentration; D: pH of medium; and E: incubation time (hour).

Glucose is one of the sugar types that could be utilized by yeast metabolism to produce bioethanol. Glucose could be produced from hydrolysis of lignocellulosic biomass by using cellulase and hemicellulase. Figure 2 illustrates the higher concentration of sugarcane bagasse did not increase the glucose concentration in the medium. Optimum sugarcane bagasse concentration in producing glucose during SSF process was 3%. This result was assumed the fermentation process in this research used the SSF system. In this fermentation, all saccharification processes and sugar metabolism are performed in one vessel. In this condition, glucose was released as a result of sugarcane bagasse saccharification directly metabolized by BTCC 3 yeast strain to produce bioethanol. Thus, glucose concentration was not detected in the fermentation culture sample. In addition, a high concentration of biomass could inhibit enzyme to hydrolyse substrate. The product of hydrolysis of biomass such as glucose and short chain of cellulose may have inhibited cellulase activity [11].

The ANOVA results for the glucose production model were not significant and the regression values were also low. The incompatibility of this model was reinforced by the optimum conditions between the actual and predictions.

3.3. Xylose

The p-value for the experimental model of xylose concentration as a fermentation product from sugarcane bagasse by BTCC 3 yeast strain was <0.0001. This value indicates that the model was significant. The F value also shows an established model. The F value exceeds the tabulated value of F at the 0.0001 probability level, indicates the null hypothesis (H0) was rejected at a significance level of 0.0001.

Experimental data and predictive xylose concentrations are given in Table 3. The ANOVA results show a quadratic regression value (R^2) of 0.9711. This value means that 97.11% of xylose concentration formed was influenced by SSF. The parameter of optimum conditions for bioethanol production from the experiments and prediction was similar (**Table 3**). The highest xylose production was 1.44 g/L at 5% sugarcane bagasse, a combination of cellulase and hemicellulase with 1 FPU/g of sugarcane bagasse, medium pH of 6.0, and after 72 hours of incubation period.

Table 3. Optimum conditions for xylose concentration with SSF by using BTCC3 yeast strain.

A Biomass concentration (%)	B Type of enzyme	C Concentration of enzyme (FPU)	D pH	E Time of incubation (hour)	Xylose (g/L)		Error (%)
					Prediction	Actual	
5	3 (Cellulase + Hemicellulase	1	6	72	1.37	1.44± 0.000	0.000

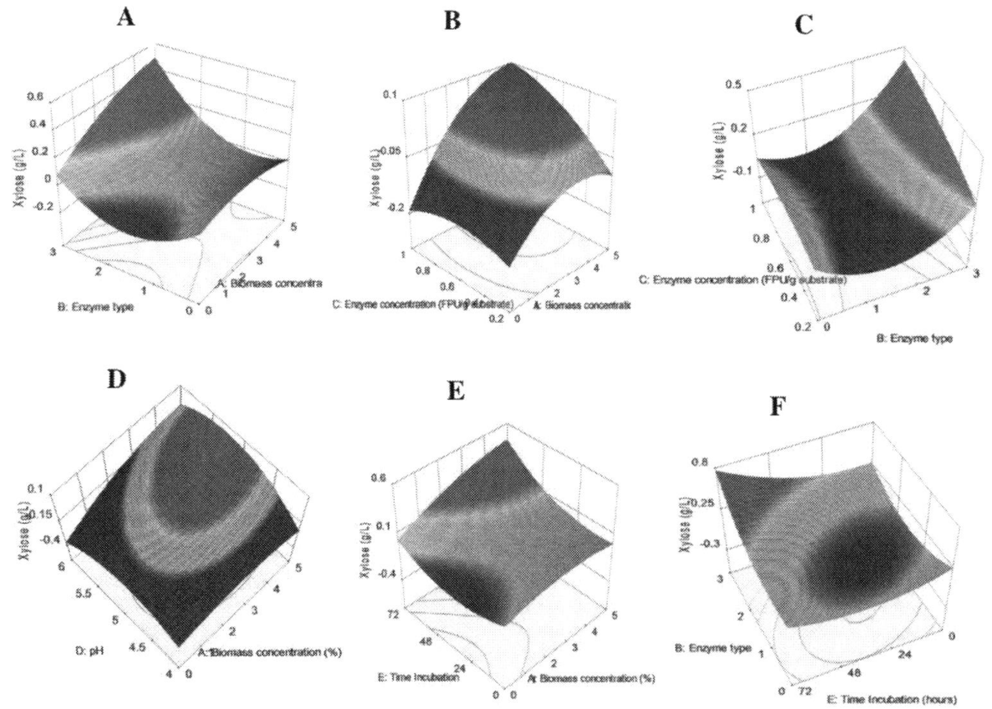

Figure 2. Xylose response in enzyme type (A), bagasse concentration (B), enzyme concentration (C), medium pH (D), incubation time and biomass concentration (E), and enzyme type and fermentation time (F). Fermentation was performed at 30 °C and agitation at 150 rpm.

The result of quadratic model application for xylose concentration gives the model as follows:

$$Xylose = + 0.032 + 0.11A + 0.15B + 0.052C + 0.029D + 0.12E + 0.11AB + 0.051AC + 0.072AD + 0.087AE + 0.076BC + 0.030BD + 0.13BE + 0.10CD + 0.064CE + 0.066DE - 0.069A2 + 0.22B2 - 0.036C2 - 0.16D2 + 0.15E2$$

Description: A: composition of bagasse (%); B: type of enzyme; C: enzyme concentration; D: pH medium; and E: incubation time.

Production xylose increased along the enhancement of sugarcane bagasse concentration, enzyme concentration, and time fermentation (**Figure 2**). Hemicellulose sugarcane bagasse was not used in yeast metabolism before hydrolysis it to sugar monomer. ANOVA results for xylose concentration in the medium had a high F-value and low Prob> F value <0.005. Thus, the predicted equation for the xylose concentration in the medium could be used for sugarcane bagasse [26].

4. Conclusions

The response surface method (RSM) was used to investigate the effects of bioethanol fermentation parameters, such as sugarcane bagasse concentration, enzyme type, enzyme concentration, medium pH, and fermentation duration. Bioethanol production was used by simultaneous saccharification and fermentation (SSF) of sugarcane bagasse hydrolysate by using *S. cerevisiae* BTCC 3. Maximum bioethanol production using SSF method was the same as RSM prediction. The optimal conditions were 5% (w/v) of sugarcane bagasse, a mixture of cellulase and hemicellulose at 1 FPU/g of sugarcane bagasse for each enzyme, pH 6.0, and 72 hours of fermentation. The bioethanol yield of 2.43 g/L was obtained under these conditions.

5. Acknowledgements

This study was supported by the International Joint Program, Science and Technology Research Partnership for Sustainable Development (SATREPS): "Innovative Bio-Production in Indonesia (iBioI)" from the Japan Science and Technology Agency and the Japan International Cooperation Agency (JST and JICA), and DIPA of Indonesian Institute of Sciences (LIPI) 2016-2017.

6. References

[1] Davila-Gomez F J, Chuck-Hernandez C, Perez-Carrillo E, Rooney W L, and Serna-Saldivar S O 2011 *Indust. Crops and Prod.* **33** 611
[2] Hahn-Hagerdal B, Galbe M, Gorwa-Grauslund M F, Liden G and Zacchi G 2006 *Trends in Biotechnol.* **24** 549
[3] Gnansounou E and Dauriat A 2010 *Bioresource Technol.* **101** 4980
[4] Wang LJ, Luo ZL, Xiu SN and Shahbazi A 2011 *Energy Sources* **33** 1230
[5] Uncu ON and Cekmecelioglu D 2011 *Waste Manage* **31** 636
[6] Sendelius J 2005 Master of Science Thesis Lund University, Sweden
[7] Quintero J A, Montoya M I, Sánchez O J, Giraldo O H and Cardona C A 2008 *Energy* **33** 385
[8] Cardona C A and Sánchez OJ 2007 *Bioresource Technol.* **98** 2415
[9] Cheng K K, Cai B Y, Zhang J A, Ling H Z, Zhou Y J, Ge J P and Xu J M 2008 *Biochem. Engin. J.* **38** 105
[10] Saxena R C, Adhikari D K and Goyal H B 2009 *Renew. and Sustain. Energy Rev.* **13** 167
[11] Lin Y and Tanaka S 2006 *Applied Microbiol. Biotechnol.* **69** 627
[12] Sánchez O J and Cardona C A 2008 *Bioresource Technol.* **99** 5270
[13] Martín C, Klinke H B and Thomsen A B 2007 *Enzyme Microb. Technol.* **40** 426
[14] Restuti D and Michaelowa A 2007 *Energy Policy* **35** 3952
[15] Cardona C A, Quintero J A and Paz I C 2010 *Bioresource Technol.* 101 4754

IOP Conf. Series: Earth and Environmental Science **183** (2018) 012010 doi:10.1088/1755-1315/183/1/012010

[16] Shen F, Saddler JN, Liu R, Lin L, Deng S, Zhang Y, Yang G, Xiao H and Li Y 2011 *Carbohydrate Polymers* **86** 1542
[17] Ohgren K, Rudolf A, Galbe M and Zacchi G 2006 *Biomass and Bioenergy* **30** 863
[18] Vanderghem C, Brostaux Y, Jacquet N, Blecker C and Paquot M 2011 *Indust. Crops and Prod.* **35** 280
[19] Lee H V, Hamid S B A and Zain S K 2014 *The Sci. World J.* http://dx.doi.org/10.1155/2014/631013
[20] Van Dyk J S and Pletschke B I 2012 *Biotechnol. Adv.* **30** 1458
[21] Wang L J, Luo Z L, Xiu S N and Shahbazi A 2011 *Energy Sources* **33** 1230
[22] Dasgupta D, Suman S K, Pandey D, Ghosh D, Khan R, Agrawal D, Jain R K, Vadde V T and Adhikari D K 2013 *SpringerPlus* **2** 159
[23] Pandey A, C R Soccol, P Nigam and V T Soccol 2000 *Biores. Techno.* **74** 69
[24] Sun J X, X F Sun, H Zhao and R C Sun 2004 *Polymer Degradation and Stability* **84** 3319
[25] Stanley D, Bandara A, Fraser S, Chambers P J and Stanley G A 2010 *J Appli. Microb.***109** 13
[26] Yingling B, C Li, Y Zongcheng and W Honglin 2011 *Antonie van Leeuwenhoek* **99** 329
[27] Fakruddin M D, Quayum M D. A, Ahmed M M and Choudhury N 2012 *Biotechnology* **11** 248

IOP Conf. Series: Earth and Environmental Science **183** (2018) 012011 doi:10.1088/1755-1315/183/1/012011

Bio-organo-chemical fertilizers: a new prospecting technology for improving fertilizer use efficiency (FUE)

Didiek Hadjar Goenadi[1*], A B Mustafa[1] and L P Santi[1]

[1]Indonesian Research Institute for Biotechnology and Bioindustry, Jalan Taman Kencana No.1, Bogor 16128, Indonesia

*E-mail: dhg_rpn@yahoo.com

Abstract. Fertilizer use efficiency (FUE) has been the focus of agriculture cultivation practices to meet economic and environmental challenges. Unfortunately, available technology for improving FUE is somewhat impotent as no significant breakthrough in fertilizer technology during the last several decades. However, by harnessing the most active components of chemical, organic, and biological sources to produce a multigroup fertilizer so-called bio-organo-chemical fertilizer, it is possible to develop a new technology employing new paradigm in fertilization, i.e. improving soil capacity. Integration of microbes as consortium packages with the recent coating technology would be a promising approach to support sustainable agriculture. Intended targets are to reduce the volume of fertilizer application and consequently to lower cost and give more profits to the farmers. The microbial coating on chemical fertilizer technique has been invented by using epigenetic approach. If this technique can be employed in addition to currently available bio-encapsulation or carrier-based techniques, then the eminent consortium of microbes in chemical fertilizer formulation will become more efficient in the future. This paper reviews new technology development of a prospective bio-organo-chemical fertilizer to enhance FUE.

1. Introduction

Soil fertility is a manageable soil property and its management is of utmost importance for optimizing crop nutrition on both a short-term and a long-term basis to achieve sustainable crop production. Soil productivity is the ability of a soil to support crop production determined by the entire spectrum of its physical, chemical and biological attributes. Current management of soils focuses primarily on maximizing agricultural production. However, long-continued use of the major nutrients nitrogen, phosphorus and potassium can bring about nutrient imbalances in soil with effects, other than toxicity, that is deleterious to crop growth [1]. Fertilizer has a significant role in increasing agricultural production, productivity, farmer income, and national food security, the government highly concerns in managing the procurement and distribution of fertilizer to the farmers. To increase fertilizer use efficiency (FUE) and to minimize its negative impact on the environment have been the focal points in the world agriculture for a long time [2].

Soil productivity encompasses soil fertility plus all the other factors affecting plant growth, including soil management. Soil fertility connotes primarily the combined effect of chemical and biological properties and is probably the most important single soil factor affecting productivity. It is very important to increase FUE, since it affects farmer's profit, especially in Indonesia whose

IOP Conf. Series: Earth and Environmental Science 183 (2018) 012011 doi:10.1088/1755-1315/183/1/012011

relatively low FUE and serious losses of nutrients. Fertilizers are considered as efficient when maximum yield is obtained with a minimum possible amount of fertilizer application. It is indeed difficult to quantify the efficiency of a particular fertilizer since it depends on (1) losses due to leaching, (2) losses in gaseous forms, (3) immobilization by chemical precipitation, adsorption on exchange complex and microbial cells, (4) chemical reactions between various components in fertilizers during mixing, before application to soil, (5) physical properties of soil, (6) chemical properties of soil and, (7) fertilizer characteristics. Completely, soil factors have a large influence on the transformation, fixation (adsorption) and leaching losses of N, P and K. Among the soil factors are its texture, proportion and amounts of clay (expanding, non-expanding and amorphous material), organic matter content, the cation exchange capacity, the concentration of ions on the exchange complex, the capacity of soil to release or renew the levels of exchangeable ion or to fix ions, soil pH, soil moisture, soil temperature, soil aeration and soil compaction In selecting the fertilizer care should be taken to select such a type of fertilizer which will have minimum interaction with soil and the time and mode of application should be such as to ensure minimum immobilization of nutrients contained in the fertilizers [3].

According to Kassam and Friedrich [4], the utilization of external inputs of mineral nutrients and chemical pest control combined intensive tillage for crop production has been very successful in the sense that it has led to the intensification of agriculture to feed the world and avoid the Malthusian outcome. But this unprecedented success with intensive tillage-based production systems has come with a large environmental and human health costs and has not brought benefits to farm households around the world. As discussed by Anas [5] the significant impact of continues and irrational application of agrochemicals on agriculture and avoidance of using organic fertilizer since the years of the seventies have been noticed significantly in Indonesia. Soil degradation, environmental pollution, levelling-off and decrease in land productivity are among the negative impacts of long application of chemical fertilizers. In the last few years, the tremendous increase in the price of chemical fertilizer has stimulated scientists, governments and farmers to seek alternative fertilizers to chemical fertilizer. Taking the example of consumption of 21 million tons of nitrogen each year, and 45% of the amount is lost. The total loss of nitrogen can reach 9.45 million tons each year, which equals 20.5 million tons of urea [2].

The heavy loss of fertilizer has triggered a series of environmental problems. In some intensified agricultural areas in the north, the irrational application of nitrogen has led to the overrun of nitrates in the groundwater. Besides, there are other examples of environmental problems resulting from the irrational application of fertilizer, such as the accumulation of nitrates in vegetables, the increased emission of nitrous oxide in the air [31]. These problems lead to a decline in soil fertility. Soil fertility decline occurs when the quantities of nutrients removed from the soil in harvested products exceed the quantities of nutrients being applied. In this situation, the nutrient requirements of the crop are met from soil reserves until these reserves cannot meet crop demands. This results in a reduction of plant growth and yield. In improving fertility decline, the enhancing of soil organic matter and microbe is a necessity. The organic matter in soil not only supplies different nutrient elements but also improves physical conditions of soils, stimulates microbial activity, protects the soil from erosion, retards the fixation of nutrients, increases the buffering capacity, and helps in many other ways. In addition, as organic matter decomposes, it releases nutrients that are bound in the organic matter's structure, essentially the ultimate slow release fertilizer. These benefits, in turn, increase the efficiency of applied fertilizers. Organic matter can also hold large amounts of water, which helps nutrients to move from soil to plant roots.

Less and less profits are enjoyed by farmers as the ever-increasing price and dosage of fertilizers, and a relatively stagnant commodity price continue. Those tremendous cases are now alarming in several parts of Indonesia or other developing countries. While farmers sell their product to the local market, the prevention or limitation mechanism of import product must be administered to help and accommodate local farmer. In the future, climate change issues could trigger the agricultural price significantly. In contrast, currently available technology for improving FUE is somewhat impotent as no significant breakthrough in fertilizer technology during the last several decades. Therefore, to solving the FUE problem, the basic knowledge formulation both chemical and biology fertilizer are highly important. There is a tendency, however, to use one single product for all, which become more efficient in time and workforce. The urgency to formulate high efficient fertilizer with combine chemical, organic material, and microbe is considerably high. The current technology available is limited in a combination of chemical and organic fertilizer (organo-chemical fertilizer) and organic and microbes (bio-organo-fertilizer). In spite of this approach, the technology combining live microbe with chemical fertilizer is unavailable today. In the future, it is strongly expected to have a more simple, accurate, and end-user friendly products. Combination of the fertilizer design and microbial mechanism allow us to the possibility of creating bio-organo-chemical fertilizer. This paper aimed to reviews currently available technology and in relation to FUE and explore a new approach to fertilization by using a new paradigm, i.e. improving soil capacity to enhance FUE.

2. Challenges in Fertilization

In reality, it is extremely unlikely that the farmers from developing countries are making profit because most farmers face economic constraints that limit their effective demand for fertilizer, technical constraints that make it difficult to use fertilizer in combination with recommended crop management practices, and institutional constraints that limit the development of human capital and markets [6]. Public spending on fertilizer subsidies has increased far more than the production costs for the industry. In the manufacture of urea, natural gas is the main input and the increasing gas costs worldwide drove international urea prices upward by 50 percent between 2007 and 2008. However, the subsidy in Indonesia grew even faster, at 142 percent over the same period, suggesting that production costs are not the sole driver of increased spending [7]. FAO [8] reported that the world economy has broadly strengthened over the past three years and is expected to continue this strengthening during 2014-2015. World consumption of the three main fertilizer nutrients, nitrogen (N), phosphorus expressed as phosphate (P_2O_5), and potassium expressed as potash (K_2O), is estimated to reach 186.67 million ton (N, P_2O_5 and K_2O) in 2016, up by 1.4 percent over 2015 consumption levels. The demand for N, P_2O_5, and K_2O is forecasted to grow annually on average by 1.5, 2.2, and 2.4 percent respectively from 2015 to 2020. Over the next five years, the global capacity of the production of fertilizers, intermediates and raw materials is also expected to increase.

Judicious use of fertilizers and lime has a number of beneficial effects on soil properties and the surrounding environment beyond the obvious enhanced food and fibre production. According to Olson [1] although claims have been made that certain fertilizers are responsible for 'hardening' soils and making them less permeable to water, the effect, were measured, usually has been the consequence of poor management by the farmer, such as too much traffic over the soil surface or pulling anhydrous ammonia shanks through the soil when too wet. Extended use of nitrogen and phosphate fertilizers, especially the former, can, however, cause an acidification problem. As a general axiom, each kilogram of fertilizer nitrogen (other than sodium and potassium nitrates) requires about 1.8 kg of lime for its neutralization, while ammonium sulphate requires three times more.

Overloading the soil with phosphorus, especially in neutral to alkaline soils, can induce zinc deficiencies. When this problem is corrected by building up soil zinc levels with fertilization, an iron shortage may develop owing to the high phosphorus and zinc levels. A problem is thus created for which economic soil treatment measures do not presently exist. Similarly, loading soils excessively with potassium and nitrogen has been found to cause low magnesium levels in forages to the particular detriment of animals consuming forage [1]. The active ingredients in chemical fertilizers are highly concentrated and are provided to the plant too quickly, with the result that much of the fertilizer is not taken up and used efficiently; the excess is absorbed by the soil, often causing such common secondary effects as "nitrogen burn" due to dehydration, as well as contamination of groundwater.

Fertilizer's agronomic potential is often unrealistic because of poor land and crop husbandry practices at the farm level. The failure to transmit research results about increasing fertilizer use efficiency (fertilizer/crop rotation interactions, use of micro-doses, soil type/fertilizer interactions) to farmers is also a factor. Many "poor" management practices (late application or inadequate doses) often stem from farmers' efforts to reduce risk. Response farming techniques and simulation models show promise for better risk management, but researchers need to transfer the lessons learned from these tools to many more farmers [6]. Rachman and Sudaryanto [9] added that national fertilizer production tends to be stagnant at an average of 75 percent utilization of the capacity. In growing markets today, some efforts to increase production capacity of fertilizer have been done through rehabilitation of urea factory in order to operate more efficiently. Up to now, the Ministry of Agriculture of Indonesia had listed 976 brands of fertilizer that consists of 296 brands of NPK Fertilizer and 407 brands of mix fertilizer. While organic fertilizer produced by the farmers reached 157000 tons. This diverse brands of fertilizer and wide product differentiation create a farmer's unfavourable situation.

3. Current Status of Available Technology

3.1 Chemical compound fertilizer

There are many different analyses of compound fertilizers available throughout the world. These are traditionally supplied from one area/source, but increasingly are becoming available from other areas. Some products of this type are also manufactured by 'blenders' where the various components are simply mixed together in the appropriate proportions, as opposed to processes involving chemical reactions or wet granulation. Most ordinary compound grades can be supplied in bulk or bags. According to FAO [10] two grades, in particular, 15:15:15 and 20:20:0, are manufactured internationally in very large volume and the market for these behaves very much as a commodity market where supply and demand predominate as an influence. Reducing the volume and increasing the ability of fertilizer consequently to lower cost and give more profits to the farmers. Volume reduction could be achieved by using a controlled release fertilizer.

Liu *et al.* [11] reported that slow-release fertilizers generally have a slower release rate of the nutrient than conventional water-soluble fertilizers and controlled-release fertilizers. Slow-release fertilizers can occasionally be released very quickly when excessive moisture and high temperatures occur in the same period of time. Controlled-release fertilizers contain a plant nutrient in a form that delays its availability for plant uptake and use after application, or that extends its availability to the plant significantly longer than "rapidly available fertilizers" such as ammonium nitrate or urea, ammonium phosphate, and potassium chloride. To match crop nutrient requirements, the ideal fertilizer should have this characteristic: the nutrient release matches the nutrient requirements of the crop throughout all of the plant growth stages. Fortunately, using deliberate applications of controlled-release fertilizers and slow-release fertilizers in specific circumstances where they are appropriate can accommodate timely plant nutrient demand requirements, maximize nutrient use efficiency, and minimize environmental concerns. In 1992, Goenadi [12,13,14,15] reported that slow release fertilizer is beneficial for estate crop seedlings (cocoa seedlings, oil palm seedlings, rubber seedlings) and showed the potential release period of the fertilizer is approximately 32 months. The newest technology in 2016 of an environment-friendly slow release fertilizer of zeolite-based nano-composite has been studied by Lateef *et al.* [16]. The zeolite-based nano-composite (ZNC) was synthesized by simple impregnation of nutrient. While Zhou *et al.* [17] developed a research of controlled-release urea fertilizers with synthesis composite hydrogels from inorganic-organic hybrids technique. Coating large urea particles with polyethylene (PE) membranes for controlled-release fertilizers also has been conducted by Yang *et al.* [18]. This type of fertilizer offers significant reductions both in volume and cost as its consequence.

3.2 Organochemical fertilizer

Organic products are eco-friendly natural sources, which can be considered as an alternative to sustainable agriculture development. Chen [19] reported that the effects of organic fertilization and the combined use of chemical and organic fertilizer on crop growth and soil fertility depend on the application rates and the nature of fertilizers used. In general, the application rates of organic fertilizer mostly are based on crop N need and estimated rates of organic fertilizer N supply but do not consider the amount of P and K provided with organic fertilizer. However, the N/P ratio of organic fertilizer usually is significantly lower than the N/P uptake ratio of the crop. Therefore, basic organic fertilizer on N supply typically results in P addition in excess of the crop's need. Nutrients, salt or heavy metal accumulation has also been reported in many papers, especially for the long-term or heavy use of organic fertilizers with higher contents of P, K, salt, or heavy metals. However, while the use of chemical fertilizers is defective in the areas of cost, scarcity and toxic residual effects, the use of organic manure becomes imperfect in terms of time and labour required for preparation, as well as the huge quantities needed to meet crop nutritional needs.

Organic fertilizers are still used in a bulky volume. Taking the most active component of organic materials called humic substance is very important. When humic substance extraction is available, the disadvantages bulky characteristic of organic fertilizer will be reduced. Positive correlations between the humus content of the soil, plant yields and product quality have been published in many different scientific journals. Quality of humic substances extracted from composts is influenced by the composting ingredients and techniques.

Several approaches that have been reported by Santi and Goenadi [20] such as organo-chemical fertilizer with humic substances utilization. Fertilizers with humic substances can be used on different types of soil, as well as in technologies for improving degraded or contaminated soils and they proved effective on a wide range of cultures. Pettit [21] explained that humic substances are the components of humus and as such are high molecular weight compounds that together form the brown to black hydrophilic, molecularly flexible, polyelectrolytes called humus. The research of [20] was to developed organo-chemical fertilizer with rock phosphate and urea. The organic materials which obtained from oil palm plantations then processing by compost methods. Afterwards, composts, rock phosphate and urea are combined through granulation process resulting of 10% C-organic, 11% N, 8% P, 1% K and 4% humic acid. While N and P are originally from rock phosphate and urea. The reaction of an organic acid from compost with rock phosphate reported increasing P content. Moreover, biomass dry weights of the seedlings applied with 100 g organo-chemical fertilizer plus 10 g KCl to each seedling were significantly different compared to the standard dosage of chemical fertilizer. Another key mechanism, which maximizes fertilizer efficiency and relates to a function of humic substances, is a reduction in the toxicity and leaching of nitrogen compounds into subsoil water. Humic substances hold these major plant nutrients in a molecular form which reduces their solubility in water. These binding processes reduce leaching nitrogen into the subsoil and help prevent volatilization into the atmosphere. This organo-chemical fertilizer could be applied as a conventional fertilizer substitute.

3.3 Bio-based fertilizer

Biofertilizers are microbial inoculants consisting of living cells of micro-organism like bacteria, algae and fungi alone or combination which may help in increasing crop productivity. Biological activities are markedly enhanced by microbial interactions in the plant rhizosphere. In contrast, organic fertilizers are obtained from animal sources such as animal manure or plant sources like green manure. For example, vermicomposting is a simple biotechnological process of composting, in which certain species of earthworms are used to enhance the process of waste conversion and produce a better end product. Judy [22] added that biofertilizer contains a variety of beneficial microorganisms inside, but all microorganisms in organic fertilizer after high-temperature treatment substantially are failed to survive, which is difficult to provide beneficial soil microbes.

A number of approaches have been investigated for incorporating microbes, or microbial compositions into fertilizer compositions. The decontaminated manure and humic acid are mixed as dry ingredients with $Ca(H_2PO_4)_2$, KCl, and Urea then fed into an agglomerator. A 50% aqueous suspension of the *Bacillus* spores is sprayed-on in the agglomerator and the resulting moistened ingredients formed into prills in a revolving drying tunnel. Other approaches are coating fertilizer particle with 5-35 dry weight % biomass solid particles with a particle size below 400 microns and an oil or wax based dispersant. There are also approaches which disclose fertilizer compositions comprising decontaminated manure and *Bacillus* spores in combination with a humic acid derivative and, optionally, one or more of N, P, K compounds. The dry ingredients are first mixed then ground to 100-150 mesh. The *Bacillus* spores are sprayed-on and the resulting product prilled via a rotating drier.

Currently, the use of organic fertilizer by farmers is still low at about 21 percent. This is due to the highest retail price of organic fertilizer which is almost equal to the highest retail price of Urea fertilizer [8]. Biological fertilizer is used in very limited volume and still in separate formulation. One product may cover mineral solubilizing microbe, and the other one provides root inoculums microbe. Biofertilizers are generally applied to soil, seeds or seedlings, with or without some carriers for the microorganisms. Regardless of methods, the number of cells reaching the soil from commercial products is smaller than the existing numbers of soil or rhizosphere microorganisms; these added cells are unlikely to have a beneficial impact on the plant unless multiplication occurs. In addition, the population of introduced microorganisms will decline and be eliminated in a very short time, often days or weeks. The formulation of inocula, method of application and storage of the product are all critical to the success of a biological product. Short shelf life, lack of suitable carrier materials, susceptibility to high temperature, problems in transportation and storage are biofertilizer bottlenecks that still need to be solved in order to obtain effective inoculation [30].

More organic matter fertilizer means also more nourishment for bacteria, fungi and other organisms in the soil. More active soil microorganisms also reduce pest populations in the soil. Some mechanistic evidence now proving that microbial consortium is key to reveal and solve not only nutrient deficiency but also protection from pest and disease. The idea to compile many broad spectra of microbial consortium package in one product now is undeniable. Even improving the chemical aspect is also possible to tackle. In the near future, one-stop fertilizer product can be distributed to the user, especially to the farmers. The benefit is also to enhance FUE by reducing chemical fertilizer dosage.

4. New Approach

4.1 Liquid vs carrier-based biofertilizer

If a scientist can integrate microbes as consortium packages with the recent coating technology, it would be a promising technology for sustainable agriculture. A preparation to preserve organisms, to deliver them to the target regions, to improve their biological activity of a consortium of microorganisms provided with a suitable medium, to keep up their viability for certain period will help in enhancing the biological activity on the target site. Pindi and Satyanarayana [23] analysed formulations comprise inoculums in semi-dry condition, includes dust, granules and briquettes, based on the particle or aggregate size. This group also comprises wettable powders formulated in dry powder. These powders require liquid or water as carrier just before the application to activate the inoculum. Dust usually range from 5-20 mm in size and require low absorbance inert diluents as carriers. Low dust particles, i.e. less than 10 mm size adhere best on the target site but are hazardous when inhaled. About 30% of the dry weight of dust contains organisms as suspension. They are normally prepared by feeding the organism into an air stream for mixing with a blender. Particle size, bulk density and flowability are extremely important.

Pindi and Satyanarayana [23] also reported the pros and cons of liquid biofertilizer (LBF) and carrier-based biofertilizers (CBBF). Liquid biofertilizer formulation is the promising and updated technology of the conventional carrier-based production technology which in spite of many advantages over the agrochemicals, left a considerable dispute among the farmer community in terms of several reasons major being the viability of the organism. Shelf life is the first and foremost problem of the CBBF which is up to 3 months and it does not retain throughout the crop cycle, LBF, on the other hand, facilitates the long survival of the organism by providing the suitable medium which is sufficient for the entire crop cycle. CBBF are not so tolerant to the temperature which is mostly unpredictable and uncertain in the crop fields while temperature tolerance is the other advantage of the liquid biofertilizers. LBF may have a better effectiveness than the solid one. The range of possible contamination is very high as bulk sterilization does not provide the desired results in the case of CBBF, whereas the contamination can be controlled constructively by means of proper sterilization techniques and maintenance of intensive hygiene conditions by appropriate quality control measures in the case of LBF. Moisture retaining capacity of the CBBF is very low which does not allow the organism viable for a longer period and the LBF facilitates the enhanced viability of the organism.

Goenadi [24] with Saraswati invented a product namely *EMAS* (Enhancing Microbial Activities in the Soils) fertilizer and conducted a research from 1992 to 1999. *EMAS* fertilizer is granular-shape biofertilizer first in the world (Patent No ID 0000294S in 1998) and significantly proved to enhance FUE up to 25% by saving chemical fertilizer dosage and increasing 10% to 30% crops yield depend on crops variety. Granulation is made by coating the granular core used a rotary-spraying drum mixer. *EMAS* fertilizer has several multiplier effect benefits such as reducing handling and transport cost, reducing environmental pollution, and increasing farmer income. Goenadi [24] stated that *EMAS* could save up to 50% conventional NPK fertilizer for soybean, chilli, and sugarcane. This wide-spectrum effect is caused by high quality and a pure isolate of microbes. Goenadi *et al.* [25] reported that an efficient bio-fertilizer product i.e. *EMAS* has been successful developed by using active ingredients consisting of *Azospirillum lipoferum, Azotobacter beijerinckii, Aeromonas punctata,* and *Aspergillus niger* native of tropical soils. The application of 6.25 g *EMAS* per plant (equivalent with 83.125 kg/ha) + 50% inorganic fertilizer recommended dosage for tea, could reduce the application of inorganic fertilizer dosage by 50% and resulted in the growth of the plant which was better than that of inorganic fertilizer. In addition, evaluation of the effectiveness of *EMAS* biofertilizer in corn at Pelaihari, South Kalimantan indicated that based on the current production value and a total of cost production, reducing 25, 50 and 75% conventional fertilizer provides the planters with 1.44, 1.13, and 1.12 revenue-cost ratio. The yield of a dry grain of corn was higher (+41.8%) by application of 75% standard dosage and 1 g *EMAS* biofertilizer/tree (53.3 kg/ha) than by standard dosage of conventional fertilizer. EMAS biofertilizer's field experiments also were conducted in plantation crops (tea, rubber, cocoa, oil palm, and sugarcane), food crops (paddy and corn), horticulture (potatoes), and spice. In addition, introducing the biofertilizer in soil provides a more stable soil aggregate and improves its fertility status.

Carrier-based biofertilizers have already proven to be the best over the agrochemicals and have been showing the tremendous effect on the global agricultural productivity for the past two decades. Rectifying the disadvantages of the carrier based biofertilizers, liquid biofertilizers have been developed which would be the only alternative for the cost-effective sustainable agriculture. Carrier and coating material are inseparable matters on the next research of bio-organic chemical fertilizer. Schoebits *et al.* [26] reported that the promising research should be based on what the carrier and how to coat it. The carrier is usually a convenient and economical material, which is able to release slowly viable cells in high-quality physiological condition, containing one or more beneficial bacteria. There are a number of alternative carriers and formulations for delivery of bacteria: talc, vermiculite, perlite, polyacrylamide, carrageenan, alginate, alginate–starch, alginate–humic acid, and powder formulations, although the most practical carriers used worldwide on commercial crops are peats, liquids, and clays.

4.2 Encapsulation technique

The state-of-the-art knowledge on bioencapsulation seems as a protective coating layer for microbes even with the mixture of chemical fertilizer one. According to Schoebits *et al.* [23], the principle of rhizobacteria bioencapsulation is to protect the microorganisms introduced into the soil and to ensure a gradual and prolonged release. The degradation rate of the encapsulation matrix will have a direct relationship with the biological activity of the soil microorganisms. The dried capsules can be stored at room temperature for a long period presenting a favourable environment for bacteria and reducing the risk of decreased survival. These inoculants can be improved by incorporating essential nutrients for bacterial growth, transforming the capsules in bioreactors, which are capable of increasing the number of encapsulated bacteria inoculated into the soil.

Schoebitz *et al.* [26] reported that bacterial inoculants have solved many problems associated with traditional peat inoculants, which originate great variability in peat quality. Numerous advantages related to the bioencapsulation of rhizobacteria are found, for instance, controlled release of bacteria into the soil, protection of microorganisms in the soil against biotic and abiotic stresses, and contamination reduction during storage and transport. Instead, bioencapsulation provides a niche where rhizobacteria are protected from the soil stress. Furthermore, the liquid inoculum, after being introduced into the soil has an instantaneous and very fast release and the rhizobacteria are delivered only in the initial moment of the plant growth. Alternatively, the encapsulated inoculants confer a gradual cell release that achieves long-term fertilizing effects. Agronomically this approach is very economical and has the lower cost compared with conventional fertilizer. Intended targets are to reduce the volume of application and consequently to lower cost and give more profits to the farmers.

The inconvenience of bioencapsulation technology must not be ignored. Many of the bioencapsulation devices are confectioned to produce beads in laboratory conditions and very small scale. This allows testing encapsulated formulations only in growth chambers and greenhouses conditions. To produce a large amount of inoculant, trials on large fields are required to use innovative bioencapsulation devices [26].

4.3 Bio-enrichment of chemical fertilizer

The urgency to enriched chemical fertilizer with the organic and biological material is an unavoidable matter. An economically feasible and environmentally product of fertilizer is desirable. Reveal the pathways of microbial life and adaptation mechanism to extreme conditions is the solution. The epigenetic approach provides an imminent path of fertilizer development. Especially in integrating microbes onto chemical and/or organic fertilizer. Epigenetic processes control central genomic functions such as the utilization of genetic information over the course of life. Epigenetic processes are controlled by adding and removing epigenetic modifications on the genes. Epigenetic is different with transgenic. Epigenetic modifications are added at different molecular levels and form a complex combination of positively and negatively regulating molecular signals. Epigenetics describes mechanisms that lead to change, heritable structural and activation states of the chromatin without changes to the primary nucleotide sequence. This molecular genetic definition describes some characteristics but leaves out some aspects of the functional consequences and other possible levels of epigenetic control. Epigenetic mechanisms are located at several levels. On the genome, the levels are those of DNA modifications and chromatin. Partly decoupled from the genome there are modifying proteins and non-coding RNA, whose site of action is in the nucleus or the cytoplasm. The common property of all three of these levels of the epigenetic mechanism is that they influence the function and regulation of genes in a long-lasting but at the same time reversible manner [27]. By understanding the potential strain of microorganisms, the more effective results can be obtained through the interaction with another active component. Soil microbiology no doubt holds the keys to unlocking some of the important new opportunities for enhancing nutrient management and crop production.

Harnessing the most active components of chemical, organic, and biological sources to produce a multigroup fertilizer is the most promising solution nowadays. One formulation called bio-organo-chemical fertilizer can be developed by understanding the characteristics of each fertilizer above and the approach is to combine them without reducing its advantageous. The combination of microbial as coating agent of granular chemical fertilizer and its effects towards plant and soil ecosystem should be explored and researched more. Schoebitz *et al.* [26] reported that the success of using microbial inoculants introduced into soil requires the survival of adequate numbers of bacteria reaching suitable habitats where they can thrive. For these reasons, the formulation is a key factor in the success of microbial inoculants. Chen [19] also reported that P-based compost application with the urea supplement was the best fertilizer management strategy for the soil and would avoid the accumulation of nutrients, salt and heavy metals. Organo-zeolitic bio-fertilizer is also developed by [31] composed of organic waste and crushed zeolitic rock, containing Clinoptilolite and commonly Mordenite zeolite, functions biologically in sponsoring nitrification. Ammonium ions, provided from the degradation of the organic waste, are adsorbed to the zeolite mineral surface thus avoiding a loss to the atmosphere by volatilization. Oxidation of the ammonium ions, by soil nitrifying micro-organisms, provides major and trace element nutrients. Analysis of pore water, from substrates amended with the bio-fertilizer, has shown that its electrical conductivity is orders of magnitude higher than that of pore water from untreated substrates. This is reflected in the high ionic concentration of cations present, which covers a wide range of elements, providing essential major and beneficial trace-elements in the ionic state that are directly available for plant uptake.

New technology that has been used is the application of spores or live cells of *Bacillus laterosarus* strain CM-3 for increasing the yields of grain crops. Using its adaptability, the spores are applied to crops as an aqueous suspension obtained directly from the fermentation of the CM-3 microorganism or via re-suspension of a spray-or freeze-dried version. The use of microbial supplements added to separate products from traditional fertilizers, introduces complexity in the target applications of agronomy, reducing their attractiveness and compromising customer adaptation and dosage compliance. Additionally, the compositions of the present invention also provide plants with higher Brix index, antioxidant levels, and chlorophyll content. The invention further provides a method of producing the fertilizer product in a solid form. The composition and methods are applicable to any microbial-based treatment designed for agronomy or agricultural applications [29].

Combination technology of microbial carrier system and chemical fertilizer such as existing particle already present in the fertilizer, e.g. Urea, Diammonium Phosphate (DAP), Potassium Chloride, or a filler particle or NPK might trigger FUE more and more. Carpenter [29] reported that the ability of *Bacillus, Pseudomonas,* and *Streptomyces* as a coating for granular NPK. Binding or coating agent is required to attach the dried microorganisms to the solid particles. Another approach is by spraying the microorganism in the liquid phase while chemical fertilizer is mixing. The adaptability of this bacteria mixture coated on the carrier system in an amount from 10^6 to 10^{11} colony forming units (CFU) per gram of carrier. Carpenter [29] also developed bacteria coated carrier and drying the coated carrier which the carrier is inert solid materials. The results showed that the activity of microbes on coated urea particles with this approach show significant nitrification from the coated urea sample indicating activity of the mixed microbial system under realistic use conditions. Parallel respirometer studies confirmed significantly higher evolution of CO_2 and consumption of O_2 on the coated urea sample than on the non-microbial control consistent with microbial growth. Agronomically, this approach can improve plant productivity and keep the health of the soil.

4.4 Prospect of bio-organo-chemical fertilizer

Improvements in management and nutrient use efficiency allow productivity to grow relatively faster than the growth rate of inputs, except in regions where fertilizer is underused. The soil and plant system is very dynamic. In addition to the chemical aspects, there is a complex biological system associated with nutrients that are normally overlooked in most nutrient management planning. The interactions of the plant root, soil mineral and organic particles, and the various kinds of micro-organisms (bacteria, fungi, protozoa, etc.) all play a role in determining the availability of nutrients from the soil to the plant. As fine-tuning of the dynamic soil-plant nutrition system continues, more economic benefits of some microbial interactions may be identified, and potential for managing some of these may be developed. This could open new aspects of plant nutrition in coming years.

The prospect of bio-organo-chemical fertilizer products is considerably high in the future. Three important sources: chemicals, organics, and microbes, can be incorporated in one formula. Bio-organo-chemical fertilizer can be realized by the new approach of chemical fertilizer enriched with organic and biological agents. These processes induce an epigenetic shift in the organisms resulting in the expression of specific attributes and generating significant performance across a broad range of application conditions. This is very important in improving soil capacity and increasing FUE. Inorganic fertilizer plays a critical role in the world's food security, but the highest yields are often the result of using organic including microbial and inorganic sources together. Integrated soil fertility management (i.e. optimizing fertilizer and organic resources, along with improved genetics, and using

IOP Conf. Series: Earth and Environmental Science **183** (2018) 012011 doi:10.1088/1755-1315/183/1/012011

modern technology) through bio-organo-fertilizer is critical to optimizing food production and efficient use of plant nutrients. Using the right source, at the right rate, at the right time, and in the right place, is a basic principle of nutrient management and can be adapted to all cropping systems throughout the world to ensure productivity, profitability and environmental stewardship are optimized. Putting the right information in the hands of the right people ensures achieving efficient use of plant nutrients [30].

5. Conclusions

Growing global population demands the safe and sufficient food for the survival of humankind. Soil health has become the greatest assertion for the scientific community in this ever-growing polluted globe. Crises of agricultural land day by day, vertical increase in the cost of agriculture input technologies are leading to transitions in the farming community. In such an agro-critical scenario, a multifaced solution for different constraints in agro-industry is necessary. A new paradigm in fertilization is not just to develop new fertilizer product and improve application technique, but more importantly is to improve soil capacity in cycling the nutrients effectively as well. Efficient plant nutrition management should ensure both enhanced and sustainable agricultural production and safeguard the environment. Chemical, organic or microbial fertilizer has its advantages and disadvantages in terms of nutrient supply, soil quality and crop growth. Developing a suitable nutrient management system that integrates the use of these three kinds of fertilizers may be a challenge to reach the goal of sustainable agriculture; however, much research is still needed. Research on microbial adaptability on chemical fertilizer are strongly needed for more economic and environmentally friendly agriculture in the future.

6. References

[1] Olson R A 1987 *Land Transformation in Agriculture*. (New York: John Wiley & Sons Ltd)
[2] Yan X, Jin J Y, He P, and Liang M Z 2008 *Agri. Sci. in China* **4** 469
[3] Dhakshinamoorthy M 2013 Online courses Tamil Nadu Agricultural University https://www.slideshare.net/mdmoorthy/fertilizer-use-efficiency
[4] Kassam A and Friedrich T 2009 *7th Conservation Agriculture Conference* (Dnipropetrovsk, Ukraine)
[5] Anas I 2010 *FNCA Biofertilizer Newsletter Issue No 8*. Ministry of Education, Culture, Science and Technology Japan.
[6] Kelly V A 2006 *Factors Affecting Demand for Fertilizer in Sub-Saharan Africa* (Washington: Agriculture and Rural Development)
[7] Osorio, C G, Abriningrum D E, Armas E B and Firdaus M 2011 *Who Is Benefiting from Fertilizer Subsidies in Indonesia?* (Washington: The World Bank)
[8] FAO 2015 *World Fertilizer Trends and Outlook to 2020* (Rome: Food and Agriculture Organization of the United Nations)
[9] Rachman B and Sudaryanto T 2010 *Analisis Kebijakan Pertanian* **8** 193.
[10] FAO 1981 *Fertilizer Procurement* (Rome: Food and Agriculture Organization of the United Nations)
[11] Liu G, Lincoln Z, Yuncong L, David D, Qingren W and Monica O H. 2014. Controlled-Release and Slow-Release Fertilizers as Nutrient Management Tools. Horticultural Sciences Department UF/IFAS Extension.
[12] Goenadi D H 1992 *Menara Perkebunan* **60** 113
[13] Goenadi D H 1992 *Menara Perkebunan* **60** 119
[14] Goenadi D H 1992 *Menara Perkebunan* **60** 122
[15] Goenadi D H 1992 *Menara Perkebunan* **60** 126
[16] Lateef A, Rabia N, Nadia J, Shahzad A, Raza S, Muhammad N K and Murtaza S 2016 *Microporous and Mesoporous Materials* **232** 174

[17] Zhou T, Yan W, Sheng H and Youcai Z 2017 *Sci Tot Env* **615** 422
[18] Yang X, Rongfeng J, Yangzheng L, Yanting L J and Lia B Z 2017 *Particuology*. In press
[19] Chen J H 2006 *Land Development Department Thailand* **16** 20
[20] Santi L P and Goenadi D H 2008 *Menara Perkebunan* **76** 36
[21] Pettit R E 2004 *Organic matter, humus, humate, humic acid, fulvic acid and humin: Their importance in soil fertility and plant health* (Texas: A&M University)
[22] Judy F 2016 *What is the difference between organic fertilizer and bio fertilizer?* https://www.linkedin.com/pulse/what-difference-between-organic-fertilizer-bio-frady-judy/
[23] Pindi P K and Satyanarayana S D V 2012 *J. Biofertil. Biopest.* **124** 1
[24] Goenadi D H 2006 *Pupuk & Teknologi Pemupukan Hayati: dari Cawan Petri ke Lahan Petani* (Jakarta: John Hi-tech Idetama) p207.
[25] Goenadi D H and Santi L P 2009 *Menara Perkebunan* **77** 58
[26] Schoebitz M, Maria DL and Antonio R 2013 *Springer Verlag*. **33** 751
[27] Walter J and Anja H 2017 *Epigenetics* (Saarbrücken: Springer Fachmedien Wiesbaden)
[28] Leggo P J 2015 *Agrotechnol* **4** 1
[29] Carpenter R S 2014 Inventor; Biowish Technologies Inc. Fertilizer compositions methods of making and using same. Patent US 2014/0352376 A1
[30] Reetz H F 2016 *Fertilizers and their Efficient Use* (Paris: International Fertilizer Industry Association)
[31] Efthimiadou A, Dimitrios B, Anestis K and Froud B W 2010 *AJCS* **4** 722

IOP Conf. Series: Earth and Environmental Science **183** (2018) 012012 doi:10.1088/1755-1315/183/1/012012

Biodegradability of oil palm cellulose-based bioplastics

Isroi[1*], A Rahman[2] and K Syamsu[3]

[1]Indonesian Research Institute for Biotechnology and Bioindustry, Jalan Taman Kencana No. 1, Bogor 16128, Indonesia
[2]Program Study of Natural Resources and Environmental Management, Bogor Agricultural University, Bogor 16680, Indonesia
[3]Department of Agroindustrial Technology, Bogor Agricultural University, Bogor 16680, Indonesia

*E-mail: isroi93@gmail.com

Abstract. This study investigated biodegradation of the bioplastic composite based on oil palm cellulose in plantation soil as natural inoculum. The biodegradation rate was evaluated as carbon dioxide released during the biodegradation test. Biodegradation test of the bioplastic compared with starch-based bioplastic and conventional plastic samples were conducted in the glass jar for 45 days. The carbon dioxide generated from the biodegradation process absorbed by 0.1 N sodium hydroxide solutions. The carbon dioxide was titrated with 0.1 N HCl and using phenolphthalein followed by methyl orange as indicator. Biodegradation rate of the bioplastic was 0.1082 mg CO_2/day and total carbon dioxide for 45 days was 8.8mg. Biodegradation rate of the starch-based bioplastic was higher than the oil palm bioplastic that is 0.157 mg CO_2/day. There was no carbon dioxide released from the conventional plastic during the test. By calculation, total degradation of the bioplastic samples were 254 days and 207 days for the oil palm bioplastic and the starch-based bioplastic, respectively.

1. Introduction

The world conventional based plastic production has reached 322 million tons in 2015 [1] and accumulation of the plastic waste to the environment poses a global environmental problem. Plastic wastes discarded in the landfills are 40% of the 70 billion pound plastics annually produced. Several hundred tons of the plastic waste are annually disposed to the ocean and finally pile up in the oceanic regions [2]. Estimation of plastic waste discarding to the marine environment is one to three orders of magnitude greater than the reported mass of floating plastic debris in high-concentration ocean gyres. According to the report, Indonesia is the second biggest plastic marine debris producer in the world [3].

The conventional plastics are made from petroleum-based materials. These plastic materials are predominantly made through carbon to carbon binds which resist degradation by microorganisms. In the recent years, there has been a growing trend of public worries to the hazardous effects of the conventional plastics on the human, organism and environment. For this reason, the research and development of biodegradable bioplastics are suitable with the concept of ecologically safe sustainable industrial development.

Oil palm bioplastic has been developed in the previous research [4]. The bioplastic was a cellulose-starch based composite. Cellulose was purified and modified from oil palm empty fruit bunches

(EFB). Combined with cassava starch as biopolymer matrix and glycerol as a plasticizer, the oil palm bioplastic was successfully made into sheet and bag. Although all components in the bioplastic were biodegradable, the biodegradability of the bioplastic has not tested. This research aimed to test the biodegradability of the bioplastic in landfill soil by natural inoculum. Degradation of the bioplastic in aerobic condition by microorganism would release carbon dioxide as degradation result. The carbon dioxide was measured by titration methods.

2. Materials and Methods

2.1. Oil palm bioplastic and references plastic samples
Oil palm bioplastic was obtained from the previous research [5]. The oil palm bioplastic was a composite bioplastic containing cellulose from oil palm empty fruit bunches; cassava starch as matrix biopolymers; glycerol as plasticizer and compatibilizer. The bioplastic film was cut into small pieces about 1 cm to 1 cm square. Starch-based bioplastic and conventional plastic were used as reference material during the test. These plastic samples obtained from the local market. All the samples were cut into 1 cm to 1 cm pieces. All bioplastic and plastic samples keep in a desiccator to maintain the humidity before use.

2.2. Inoculum source
The inoculum used in the biodegradation test was a local plantation soil in Dramaga, Bogor, West Java, collected from the top soil 5 cm to 10 cm in deep. The soil was screened to remove debris and an organic material. Total plate count (TPC) of the soil was 21.3 x 10^5 colonies per gram of soil.

2.3. Biodegradation test
Biodegradation test was conducted in a 300 ml glass bottle arrangement as **Figure 1** according to the reference [6]. 100 g of the plantation soil was placed into the samples bottle. Each bioplastic and reference plastics samples (±0.1 g) were placed and buried into the soil. A bottle without bioplastic or plastic samples was used as a control test to correct the carbon dioxide released during the test. Carbon dioxide was channelled using the tube to the other bottle with 50 ml of 0.1 N sodium hydroxide. The carbon dioxide would react with the sodium hydroxide. All experiment held in Duplo experiment and at room temperature (28 °C) for 45 days. Every three days the sodium hydroxide was replaced and analysed for carbon dioxide.

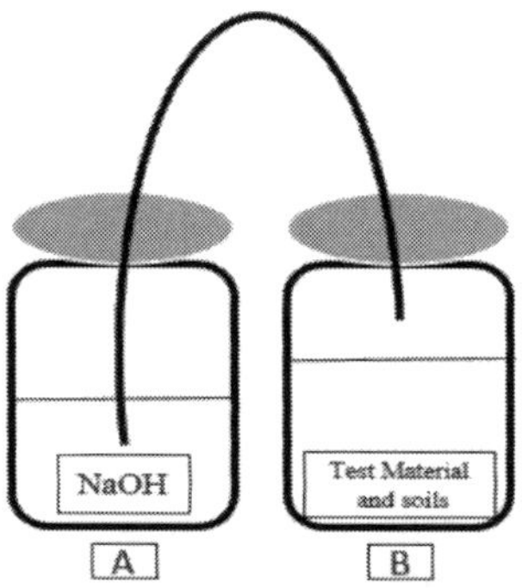

Figure 1. Experimental setup of the biodegradation test, (A) carbon dioxide trap with 0.1 N NaOH, (B) test material and plantation soil.

2.4. Carbon dioxide analysis
The carbon dioxide entrapped in the sodium hydroxide determined by titration with 0.1 N hydrochloric acid solutions. Phenolphthalein (PP) was used as a first indicator and methyl orange as the second indicator. Amount of hydrochloric acid needed to neutralize the solution into pink is proportional with the carbon dioxide releases during the biodegradation test.

3. Results and Discussion

3.1. Carbon dioxide released during the biodegradation test

Plantation soil rich in microorganisms and the total colony was 21.3 x 10^5 colonies per gram soil. In aerobic condition, the microorganism would degrade the organic materials into carbon dioxide, water and produce heat [7]. Evolution of the carbon dioxide during the biodegradation test was shown in **Figure 2**. The carbon dioxide released fluctuates. Carbon dioxide from the control experiment was carbon dioxide released from the native organic carbon source in the plantation soil. Carbon dioxide released from the samples experiments were sum of the carbon dioxide released from the samples and from the native organic carbon source.

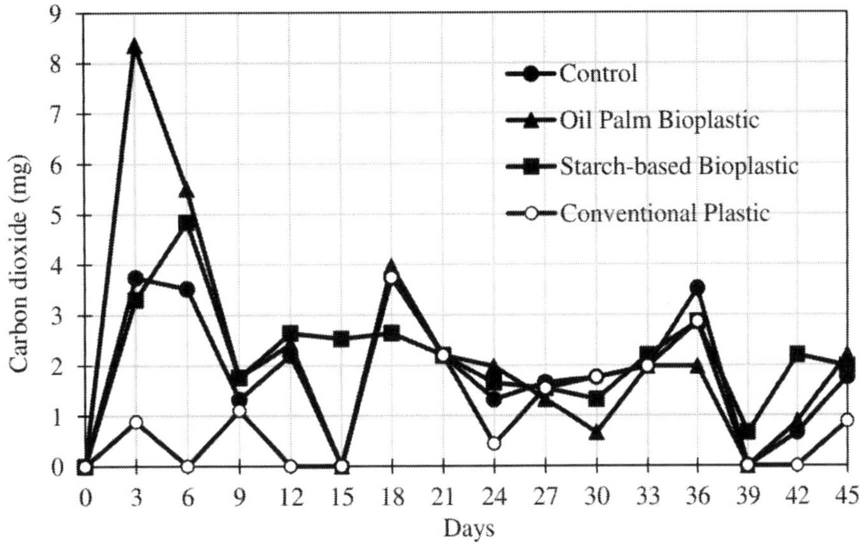

Figure 2. Carbon dioxide (mg) released during the biodegradation test.

During the first 9 days, the microorganisms were very active to decompose the organic carbon as shown by high carbon dioxide released. Evolution of the carbon dioxide gradually reduced and became more stable three weeks after incubation. Evolution of the carbon dioxide on the control and conventional plastics were lower than on the bioplastic samples. This evidence was supposed that no carbon dioxide released from the conventional plastic.

3.2. Biodegradation rate

Sum of the carbon dioxide released from the bioplastic samples and subtracted by the control experiment represent the biodegradation of the bioplastic samples. Accumulation of the carbon dioxide during the biodegradation test was shown in **Figure 3**. Carbon dioxide released in the conventional plastic was zero or negative values. The negative values were presented as zero values that mean no carbon dioxide released from the samples. In other words, no biodegradation detected in the conventional plastic experiment.

IOP Conf. Series: Earth and Environmental Science **183** (2018) 012012 doi:10.1088/1755-1315/183/1/012012

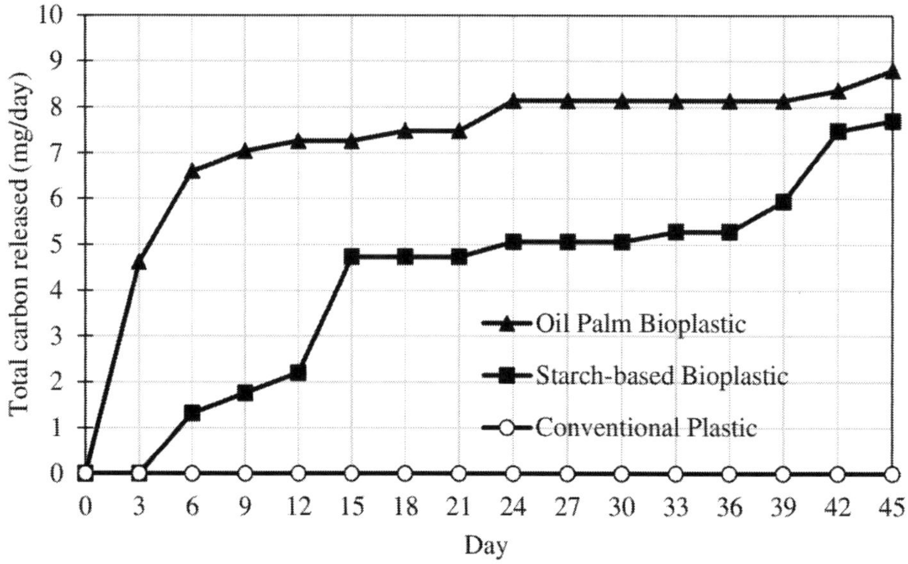

Figure 3. Total carbon dioxide (mg/day) released during the biodegradation test from the bioplastic and plastic samples corrected with the control experiment.

Biodegradation of the oil palm bioplastic was higher than the starch base bioplastic as shown in **Figure 3**. Oil palm bioplastic already biodegraded from the beginning until nine days, then the biodegradation rate slower and become stable after three weeks. The starch-based bioplastic initiated to biodegraded at the day three. The carbon dioxide released from the starch-based bioplastic was lower than from the oil palm bioplastic until 45 days.

Regression analysis of the biodegradation data could use to predict the biodegradation rate of the bioplastic samples. Biodegradation rates of the bioplastic samples were 0.1082 mg CO_2/day and 0.1568 mg CO_2/day for the oil palm bioplastic and the starch-based bioplastic, respectively. It means that the starch-based bioplastic is easier to be biodegraded by microorganisms than the oil palm bioplastic. Total carbon dioxide released during the biodegradation test was 8.80 mg CO_2 and 7.7 mg CO_2, respectively. By calculation, total degradation of the bioplastic samples was 254 days and 207 days, respectively. Resume of the biodegradation test presented in **Table 1**. The oil palm bioplastic contains cellulose derivative. The cellulose is more stable to biodegradation than starch even though the building blocks are the same. Lower biodegradation rate of the oil palm bioplastic than the starch-based bioplastic could be caused by the cellulose content in the bioplastic.

Table 1. Resume of the biodegradation test of the bioplastic samples

	Theoretical CO_2 (mg)	Biodegradation rate (mg CO_2/day)	Prediction of the total biodegradation (day)
Oil palm bioplastic	31.11	0.068	244
Starch based bioplastic	31.98	0.173	200

4. Conclusions

Advantages of the bioplastic compared to the conventional plastic are biodegradability in the natural environment. This study is successfully shown the biodegradation of the bioplastic in plantation soil. Oil palm bioplastic and starch-based bioplastic could be biodegraded by microorganisms in the plantation soil. On the other hand, conventional plastic could not be biodegraded by the microorganisms. Biodegradation rate of the oil palm bioplastic was lower than the starch base bioplastic. The lower degradation rate of the oil palm bioplastic is possibly caused by cellulose content in the bioplastic.

5. Acknowledgements

This research is funded by Oil Palm Fund Management Agency with the Oil Palm Grant Research programme in 2015.

6. References

[1] PlasticsEurope, "World Plastics Production 1950 – 2015," 2017. [Online]. Available: https://committee.iso.org/files/live/sites/tc61/files/The Plastic Industry Berlin Aug 2016 - Copy.pdf

[2] Jafari-Sales A, Shahniani A R, Bagherizadeh Y, Alizadeh S, Jahangiri-Hoseinabad M, Abdoli-Seuejani M, Bamzadeh Z and Rasi-Bonab F 2017 *Electron. J. Biol.* **13** 3

[3] Jambeck J R, Geyer R, Wilcox C, Siegler T R, Perryman M, Andrady A, Narayan R and Law K L 2015 *Sci.* **347** 768

[4] Isroi, Cifriadi A, Panji T, Wibowo N A and Syamsu K 2017 *IOP Conf. Ser. Earth Environ. Sci.* **65** 12011

[5] Isroi, Cifriadi A and Panji T 2016 Bioplastic Production from Oil Palm Empty Fruit Bunch *International Conference on Biomass 2016, 10-11th October 2016.*

[6] Khaswar S, Setyowati K and Khoiri A A 2008 *J. Teknol. Pertanian* **4** 4

[7] Adhikari D, Mukai M, Kubota K, Kai T, Kaneko N, Araki K S and Kubo M 2016 *J. Agric. Chem. Environ.* **5** 23

IOP Conf. Series: Earth and Environmental Science **183** (2018) 012013 doi:10.1088/1755-1315/183/1/012013

Introduction of bioprospecting opportunities for Indonesian mangrove species

Julkipli[1], R R Batubara[2], G E Jogia[1], I Batubara[2], K A Audah[1*] and K N Nunuk[2]

[1]Department of Biomedical Engineering, Swiss German University, The Prominence Tower, Jalan Jalur Sutera Barat Kav 15, Alam Sutera, Kota Tangerang 15143, Banten, Indonesia
[2]Pusat Studi Biofarmaka IPB (LPPM IPB), Jalan Sempur Kidul No.69, Sempur, Kota Bogor, Jawa Barat 16129, Indonesia

E-mail: kholis.audah@sgu.ac.id

Abstract. Indonesia is one of the world's most biodiverse countries. This study would be focused only on the Indonesian mangrove forests biodiversity. There are about three million hectares of mangrove forests that grow along the 95,000 km of Indonesian coastline. Mangrove forests have ecology, social, economic and medicinal values that have been used by people who live along coastal area for centuries. Many studies have shown that mangrove extracts contain many bioactive compounds that have the medicinal potential for a variety of diseases. However, mangrove plant extracts are yet to be commercially formulated as modern medicines. Although Indonesia is home to one of the largest biodiversity of plants, the interest of pharmaceutical industries in the development of herbal medicines as drugs are not as promising as those from chemical synthetics. One of the reasons for this phenomenon is a low interest in synthesizing bulks of natural products. In addition, there is a lack of facilities which can provide optimization of the herbal materials. The aim of this article is to give a rational approach for designing a bioprospecting program as an initiation on the primary screening of novel drugs from Indonesian mangrove species.

1. Introduction

Indonesia has a large number of plant biodiversity, approximately 30,000 – 50,000 species of plants, but only 7500 species of them known to be efficacious as drugs ingredient [1]. In term of mangrove biodiversity, Indonesia represents approximately 22.6% of the total mangrove ecosystems in the world [2]. Therefore, Indonesia becomes the country with the world's largest mangrove forest [3]. According to Romimohtarto and Juwana [4], there are at least 202 mangrove plant species recorded in Indonesia, including 89 types of tree, 5 types of Palma, 19 types of scandent, 44 types of herbaceous ground, 44 types of epiphyte and one type of fern. Some types of Indonesian mangrove are *Avicennia*, *Sonneratia*, *Rhizophora*, *Ceriops*, *Bruguiera*, *Lumnitzera*, *Excoecaria*, *Xylocarpus*, *Aegiceras*, *Scyphyphora*, and *Nypa* [5].

Mangrove is a plant that well adapted to many ecological stresses by alterations not only their morphology but also physiological processes resulting in the synthesis of novel chemical compounds that offer protection to these plants against various biotic and abiotic stresses. So, it should not be surprising that there are a large number of different chemical properties that exist for different functions [6]. A number of mangrove's secondary metabolites have significant pharmacological properties that have been used traditionally for the treatment of diseases [7]. Mangroves are a very potential as drugs source among a great number of wild medicinal plants biodiversity.

The prospective study of biodiversity looking for applications in agriculture, industry and medicine is known as bioprospecting. However, the best known is in the pharmaceutical industry. In this way, natural products have important contributions in drug discovery as acetylsalicylic acid (ASA) (painkiller derived from willow bark, *Salix alba*), reserpine (antihypertensive from *Rauwolfia serpentina*, known as Indian snake root), d-tubocurarine (muscle relaxant used in anesthesia from *Chondrodendron tomentosum*), artemisinin (derived from *Artemisia annua* used as an anti-malarial agent), vincristine and vinblastine (anti-cancer drugs derived from *Catharanthus roseus* [8]. Therefore, mangrove plants with numerous bioactive compounds should be suitable for bioprospecting program.

The aim of this article is to provide a rational approach to design a bioprospecting program as an initiation on the primary screening of novel drugs from Indonesian mangrove species as a following up of extract library establishment of Indonesian mangrove initiated by Audah *et al.* [9]. Hence, mangrove forests suffered a shrinkage of about 1-2% every year globally [10]. If the biodiversity is not properly managed, it will soon only become a history.

2. Materials and Methods

Bioprospecting can be achieved through information management and inventory of biodiversity, basic research, applied research, product development, production, marketing, and distribution. But in this article only a few steps will be discussed, among others:

2.1. Information management and inventory of mangrove diversity

The approach in the sample collection can be based on its use as a traditional medicine and may also be randomized. The nationally programmed collection will make it easier for research related to the conservation and understanding of its genetic resources. Identification of the species taxonomy is very important both to increase the probability of finding new species containing new compounds and avoiding known compounds.

The collection process should take into account the concept of species conservation and the habitat of collection subjects. Because research on target compounds will require more material in progressive amounts and will ultimately have an impact on the environment. Therefore, it must be ensured that the initial collection is well documented. To ensure the same samples the prior sampling process must be equipped with GPS data, complete documentation and good taxonomic knowledge [11].

2.2. Development of extraction method

Extraction must ensure that all extracts are available for screening, isolation, and elucidation of chemical structures. Degradation of compounds can occur in extracts stored for long periods of time. The natural product can be stored as simplicial which can maintain the integrity of the compound. However, humidity should be controlled to prevent damaged samples; high humidity increases the chances of fungi and microorganisms growing. Samples should be stored in each particular container and given a barcode.

2.3. Screening of antimicrobial and anticancer activity

Drugs discovery derived from plants has been the focus of world researchers since they have the potential to counter the threat of resistant bacteria and have anticancer properties. Several *in vitro* methods may be used to evaluate the antibacterial activity of extracts or pure compounds. The most well-known basic methods are the disc diffusion and broth dilution method. The screening process for potential anti-cancer natural ingredients can be performed by Brine Shrimp Lethality Test (BSLT). BSLT is considered quite economical and uses a small amount of test material. Since its introduction, *in vivo* tests have been shown to be represented as a cytotoxic active fractionation bioassay guide and anticancer agents [12].

2.4. Fractionation, isolation of pure compounds and establishment of extract library

When the extract exhibits biological activity, fractionation is necessary to separate groups of compounds with similar chemical-physical properties, such as solubility and acidity [13]. Each fraction was tested and the active sample had to be fractionated repeatedly to increase the purity of the compound. The process of testing and fractionation is carried out continuously until a pure compound is obtained which is responsible for certain biological activities. However, the repeated fractionation process needs to be prevented by the process of dereplication [14].

The extract library is a collection of active compound extracts from natural ingredients used for screening biological targets. A quality extract library will be a way of discovering natural materials that can be developed into drugs and become the identification of the starting point of chemical drug optimization.

2.5. The establishment of extract library consortium

The establishment of *extract library consortium* was initiated with the aim that the research activities of medicines from natural materials, especially mangroves, which have been done separately by each researcher, can be better organized, synergize and avoid overlap with each other.

3. Results and Discussion

The crude extracts obtained from the study were 64 extracts. The extraction was conducted by a re-maceration method with four different solvents (hexane, ethyl acetate, ethanol and water). In general, most extracts is effective against gram-positive bacteria and have potential to be an anticancer agent. Mangrove plants that used in the research are 8 species out of more than 20 species, so there are more bioactive compounds that have not been explored since different species might contain different bioactive compounds.

The crude extracts that are shown to have further activity can be fractionated and isolated to obtain a single compound which can be developed into a natural medicine ingredient. Furthermore, the single compound can be optimized to achieve safety and efficacy according to established medical and standard requirements. Preclinical testing of mammalian objects from mice, rabbits, or even primates can be performed to ensure the desired safety and efficacy. In order to compete in the world market, in addition to product quality there is also a very decisive factor, namely the genetic resources of products of interest to the world market and have access to quality databases of various commodities that become a market demand.

Qualitative phytochemical analysis showed that mangrove leaf extract of *Rhizophora stylosa* and *Avicenna marina* contained tannins, flavonoids, terpenoids, alkaloids, flavonoids, and phenolic glycosides [15]. Mangrove extract also contains several bioactive compounds such as flavonoids, saponins, tannins, and triterpenoids [16]. As far as the author's observation, mangrove utilization in Indonesia is limited to conservation only. Potential mangroves as ingredients of the drug received less attention. Therefore, bioprospecting programs for mangrove plants need to be initiated.

Essentially bioprospecting has two basic goals, among others (1) sustainable use of genetic resources and conservation, and (2) socio-economic development for a country rich in biodiversity. This modern bio-prospecting concept gives developing countries a way to improve their national capacity to add value to natural resources, build skills, infrastructure and technology to develop new products for global markets while ensuring the protection and use of natural resources that is sustainable [6]. In conducting biodiscovery research, it should be noted that if large numbers of re-collections may not be possible, so the target compounds should be chemically synthesized.

Bioprospecting is constituted as scientific and economic activities with a high impact on sustainable development and poverty reduction as well as economic growth for developing countries. Because the establishment biodiversity-based industries produce social appropriation of knowledge showing the benefits and applications of biodiversity [6]. Also, the revenue from royalties represented by the development of new medicines (have been estimated that a successful drug can generate $1 billion per year) derived from natural sources contribute to the conservation of biodiversity in mega diverse countries as well as the protection of traditional medical knowledge [8].

The main parameters to be considered in bioprospecting are the most energy and capital investment needs in the technological and product development process. For example, in the search for drugs, the most complex part is basic research of disease and potential clinical testing of drugs. The main screening of compounds is usually cheap and technically easy to do. A country wishing to promote national capacity-building in drug development can follow a capability-enhancing model that begins with a major screening for obtaining compounds and access to technology in screening, while testing and clinical trials are at an early stage [8].

4. Conclusions

Mangrove plants are potential for bioprospecting program. Bioprospecting should be based on sustainable use of biodiversity. Access to genetic resources should be undertaken taking into consideration the equitable sharing of benefits from the resulting product to be able to make genetic resources as the backbone of socio-economic development, human resource development, science and technology capability, market analysis, sustainable capital, and strategic plan should be developed.

5. Acknowledgements

The author would like to express the deepest gratitude to thank Swiss German University (SGU), ARCS Director and SGU ARCS, staff. The author would also like to thanks to lecturer's staff of FK UNILA and DKKP staff and all parties who have helped the author's research activities either directly or indirectly.

6. References

[1] LIPI 2015 Indonesia miliki 7.500 tanaman obat [Internet] Cited 2018 October 28 available from http://lipi.go.id/berita/single/Indonesia-Miliki-7500-Tanaman-Obat/11540 (*In Indonesian*)
[2] Giri C, Ochieng E, Tieszen L L, Zhu Z, Singh A, Loveland T, Masek J and Duke N 2011 *Global Ecology and Biogeography* **20** 154
[3] Giesen W, Wulffraat S, Zieren M and Scholten L 2006 *Mangrove Guide Book for Southeast Asia* (Thailand: Dharmasarn Co. Ltd)
[4] Romimohtarto K and Juwana S 2001 *Biologi Laut. Ilmu Pengetahuan Tentang Biota Laut* (Jakarta: Djambatan)
[5] Noor Y R, Khazali M and I N N S 2006 *Panduan Pengenalan Mangrove di Indonesia* (Bogor: Wetland International Indonesia Programme)
[6] Senthilkumar N and Murugesan S 2012 *Current Biotica* **5** 522
[7] Bandaranayake W M 2002 *Wetland Ecology and Management* **10** 421
[8] Juan B 2017 *J. Microb. Biochem. Technol* **9** 1
[9] Audah K A 2016 *Pros. Seminar Nasional Kimia* No B002 (Lombok: West Nusa Tenggara) p 11
[10] UNEP 2013 Mangrove forest cover fading fast [Internet] Cited 2018 October 28 available from http://www.unep. org/dewa/portals/67/pdf/GEAS_Emerging_Issues_2013.pdf
[11] WHO 2005 *Guidelines for sampling of pharmaceutical products and related materials* WHO Technical Report Series, No.929
[12] Ahmed Y, Sohrab H, Al-Reza S M, Shahidulla T F, Hasan C M and Sattar M A 2010 *Food Chem Toxicol* **48** 549
[13] Hughes J P, Rees S, Kalindjian S B and Philpott K L 2010 *British J. Pharmacol* **162** 1239
[14] Katiyar C, Gupta A, Kanjilal S and Katiyar S 2012 *Ayurveda* **33** 10
[15] Mouafi F E, Abdel-Aziz S M, Bashir A A and Fyiad A A 2014 *World Applied Sciences J* **29** 547

IOP Conf. Series: Earth and Environmental Science **183** (2018) 012013 doi:10.1088/1755-1315/183/1/012013

[16] Rohaeti E, Batubara I, Lieke A and Darusman L K 2010 *Pros Seminar Nasional Sains III* (Bogor: West java)

IOP Conf. Series: Earth and Environmental Science **183** (2018) 012014 doi:10.1088/1755-1315/183/1/012014

Effect of bio-silica on drought tolerance in plants

Laksmita Prima Santi[1*], Nurhaimi-Haris[1] and D Mulyanto[2]

[1]Indonesian Research Institute for Biotechnology and Bioindustry, Jalan Taman Kencana No.1, Bogor 16128, Indonesia
[2]Pembangunan Nasional Yogyakarta University, Jalan SWK 104, Condongcatur, Yogyakarta, Indonesia

*E-mail: laksmita.santi@gmail.com

Abstract. Drought is considered the single most devastating environmental stress, which decreases crop productivity more than any other environmental stresses. The main consequences of drought in crops are reduced rate of cell division, leaf size, stem elongation, root proliferation, disturbed stomatal oscillations, plant water and nutrient relations with diminished crop productivity. Many studies demonstrated that the deposition of silica in plant tissues can reduce drought stress. Silicon (Si) has beneficial effects on many crops, mainly under biotic and abiotic stress. Despite its abundance, silicon is never found in plants in an available form and is always combined with other elements, usually forming oxides or silicates. Monosilicic acid (H_4SiO_4) is the form of silicon used by plants, which is found both in liquid and adsorbed phases of silicon in soils. The concentration of the H_4SiO_4 in soil solution is influenced by the soil pH, clay, minerals, organic matter and Fe/Al oxides/hydroxides, which are collectively related to the geological age of the soil. Furthermore, to improve plant-available Si in the soil, silicate solubilizing microorganisms (SSM) are potentially useful in solubilizing insoluble forms of silicate. This paper aimed to review an important issue of bio-silica effects on plants under drought stress and water deficiency.

1. Introduction

Drought is one of the gravest threats to plants, and due to global warming; its prevalence is increasing worldwide [1]. Approximately one-third of the world land area is prone to drought [2]. Drought induces various changes in morphological, metabolic, or/and physiological functions of plants. At the initial phase of plant growth and establishment, it negatively affects both elongation and expansion growths [3, 4]. Reduced leaf growth and in turn the leaf areas and higher root/shoot ratio in response to drought has also been reported in many species [5]. Severe water stress poses injurious effects on plant water relations, photosynthesis, ion uptake, and nutrient metabolism and assimilates partitioning [5, 6]. Interrupted water supply from the xylem to the surrounding elongating cells under drought stress leads to loss of turgor and stomatal closure [7]. It also disturbs the photosynthetic apparatus through its interaction with UV or/and visible radiation [8]. Both stomatal and nonstomatal limitations are generally considered to be the main determinant of reduced photosynthesis under drought stress [6]. Ma and Yamaji [9] stated that the beneficial effects of Si on plant growth are commonly observed in plants under stress conditions. From a physiological point of view, for the growth and development of plants, silicon has demonstrated beneficial effects on the growth and development of plants in the increased production of various crops. Application of silicon may be a facile means to increase crop yield during drought. Moreover, because silicon can improve drought tolerance of plants, its application may help reduce the need for irrigation.

Silicon (Si) plays an important role in plant health when plants are exposed to multiple stresses. As a physicomechanical barrier, Si is part of the epidermal cell walls and vascular tissues in stems, pods, leaves and bark [10]. According to [11] and [12] indicated that Si might decrease the negative effects of oxidative stress and offer slight resistance to some abiotic and biotic plant stressors. Thus, using Si instead of herbicides and pesticides could reduce harmful environmental effects [13-15]. Some studies indicate that Si application is able to avoid the damage of the plant when grown under drought stress conditions [16-20].

Unfortunately, many soils are considered lack of available silicon that has positive impact on plants by stimulating nutrient uptake and plant photosynthesis, decrease susceptibility to disease and insect damage, alleviate water and various mineral stresses and decrease the toxic effects of aluminum and heavy metals [21-26]. Furthermore, microorganisms play an important role in the weathering of silicate minerals. We found that solubilisation of silica *in situ* in soil by silicate solubilising bacteria to supplement crop would beneficial to improving drought tolerance. This paper aimed to review an important issue of bio-silica effects on plants under drought stress and water deficiency. This may be a promising strategy for improvement of soil and plant productivity.

2. Silicon in Soil

Silicon (Si) is the second most abundant element after oxygen in the soil. Silicon dioxide comprises 50–70% of the soil mass. Therefore, all plants rooting in soil contain some Si in their tissues. However, the role of Si in plant growth and development was overlooked until the beginning of the 20th century [27, 28]. In nature, silicon occurs generally in the form of silicates, including ferromagnesian silicates (e.g. olivine, pyroxenes, and amphiboles), aluminosilicates (e.g. feldspar, mica, and clays), and silicon dioxide (e.g. amorphous silica, quartz). In general, the silicon in silicate minerals is surrounded by four oxygen atoms in tetrahedral fashion [29]. Silicon content in soil varies from 50 to 400 g Si per kg of soil [14]. In soil, Si compounds mainly present as SiO_2, about 50–70% of the soil mass, and in various aluminosilicate forms [30]. Although Si is abundant in soil, most of its sources are not available for plant uptake due to the low solubility of Si compounds in soil [31, 32, 33]. Bio cycling of silica in the soil also occurs through microbial activities that involve fungi, bacteria, and actinomycetes. Thus, plants and microbes, through their intricate interplay with soil minerals, contribute appreciably to the global silicon cycle.

3. The Function of Silicon in Plants

Silicon (Si) is an abundant element which, when supplied to plants, confers increased vigour and resistance to exogenous stresses, as well as enhanced stem mechanical strength. Plant species vary in their ability to take up and to accumulate Si under the form of silicon dioxide (SiO_2) in their tissues. Silicon as a macroelement has a vital role in plants cycles. This element is the eighth most common element in nature and the second most common element found in the soil after oxygen. One of the main functions of Si is improving the plant's growth and yield, especially in stress condition. To achieve plant tolerance, Si promotes plant photosynthesis by favourably exposing leaves to light. On the other hand, the role of the macroelement has proven to be in response to different abiotic and biotic stresses [34]. Monocots usually accumulate more Si than dicots; however, the impact of Si on dicots was greater, notably on economically important dicots. This is a subject requiring further study [35]. One major contribution of Si is a reinforcement of cell walls by deposition of solid silica. It is translocated from the roots as silicic acid [$Si(OH)_4$] through the xylem until deposition under the cuticle and in intercellular spaces. These silica bodies are called phytoliths, or plant opal. These structures are very resistant to decomposition. In addition to naturally occurring soluble Si in the soil, many crops respond positively to additions of supplemental Si.

Plants, especially grasses, can take up large amounts of Si contributing to their mechanical strength. Besides a structural role, Si helps to protect plants from insect attack, disease, and environmental stress. For some crops, Si fertilization of soils increases crop yield even in unfavourable growing conditions and in the absence of disease. The mechanism of Si in triggering a range of natural defences is through hydrolytic enzymes and some active compounds as well. The presence of Si stimulates activities of chitinase, peroxidase, polyphenol oxidases, and flavonoid phytoalexins which providing protection against fungal pathogens [22]. Regardless of the mechanism, some observed benefits due to Si nutrition include (i) direct stimulation of plant growth and yield through more growth and plant rigidity, (ii) suppression of plant diseases caused by bacteria and fungi (such as powdery mildew on cucumber, pumpkin, wheat, barley; gray leaf spot on perennial ryegrass; leaf spot on Bermuda grass; rice blast), (iii) improved insect resistance (such as suppression of stem borers, leaf spider mites, and various hoppers), (iv) alleviating various environmental stresses (including lodging, drought, temperature extremes, freezing, UV irradiation), (v) and chemical stresses (including salt, heavy metals, and nutrient imbalances).

4. Effect of Bio-Silica on Drought Tolerance

Drought is one of the most limiting environmental stresses for plant production [36]. The growth and development of plants experiencing occasionally periods of drought depend on the ability of stomata to control water loss. Plants respond to drought by closing their stomata, which reduces leaf transpiration and prevents excessive water loss in their tissues. The control of leaf stomata closure is a crucial mechanism for plants since it is essential for both CO_2 acquisition and desiccation prevention [37]. Leaf water potential and water content decrease substantially when plants are exposed to drought [6]. Application of silicon can significantly improve water status in non-irrigated crops. Based on our research, the application of silicate solubilizing bacteria (SSB-bio-silica) on oil palm seedling have a significant effect on stomatal opening in the period after drought stress treatment (**Figure 1**).

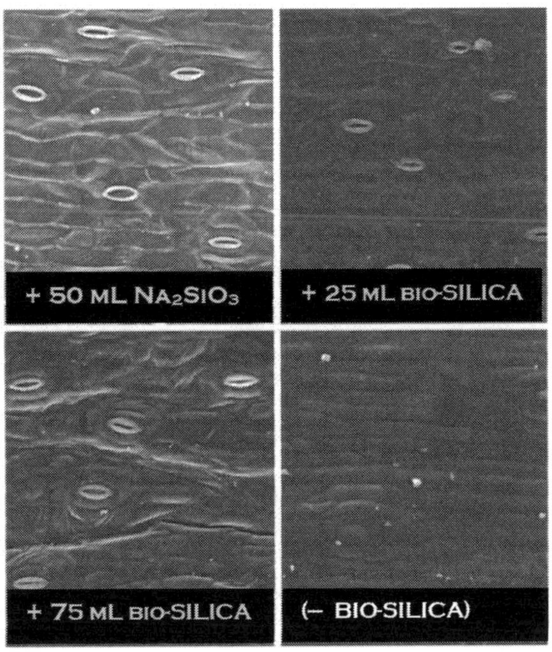

Figure 1. Scanning electron microscopy of oil palm leaf surfaces showing open stomata in application Na_2SiO_3 and Bio-Silica and closing stomata in treatment without application Bio-Silica in the period after drought stress treatment.

IOP Conf. Series: Earth and Environmental Science **183** (2018) 012014 doi:10.1088/1755-1315/183/1/012014

Silicate solubilizing bacteria (SSB) occurring in soil and rhizosphere, solubilize silica and render it available to plants. Inoculation of SSB with organosiliceous rice straw, husk and husk ash (black char/ash) to rice was found enhancing the growth, chlorophyll content, thousand grain weight, matured grains, biomass and yield [38]. The SSB inoculation along with fly ash to rice reduced the incidence of stem borer, leaf folder and gall midge and increased the yield [39]. Our study indicated that application bio-silica on oil palm seedling improved total chlorophyll in leaf up to 20% compared to control.

Microorganisms are able to degrade silicates including aluminium silicates. In the course of the metabolism in microbes, numerous organic acids are produced and have a twin position in silicate weathering. They supply H^+ ion to the medium and promote hydrolysis. Natural acids like citric, oxalic acid, keto acids, and hydroxy carbolic acids which form complexes with cations, promote their elimination and retention within the medium in a dissolved state. Santi and Goenadi [40] reported that *Burkholderia cenocepacia* KTG, *A. punctata* RJM 3020, and *B. vietnamiensis* ZEO3 have shown to accelerate the solubilization of SiO_2 originated from quartz. The amount of Si, Ca, and Mg solubilized from quartz mineral were optimum in 96 hours incubation periods. All isolates were gram-negative, a rod shape, producing exo polysaccharide, organic acid (citric, acetic, and oxalic acid), and have several different of biochemical reactions specific to each. Acetic acid was predominant substances produced by SSB isolates in liquid Bunt and Rovira media.

Several mechanisms of silicate dissolution by SSB involving acidolysis, alkaline hydrolysis, ligand degradation, enzymolysis, capsule adsorption, extracellular polysaccharides and redox play a role in the bacterial dissolution of silicate. Acidolysis is the major mechanism of weathering silicate minerals [41]. In our research, *B. cenocepacia* KTG, *A. punctata* RJM3020, and *B. vietnamiensis* ZEO3 isolates were capable of solubilizing silicates and produced acid as detected by yellow halo formation on solid Luria Bertani media containing 0.2% (v/v) bromophenol blue (**Figure 2**). As an acid-base indicator, bromophenol blue useful range lies between pH 3.0 and 4.6. It changes from yellow at pH 3.0 to blue at pH 4.6. The colour change on solid Luria Bertani medium amended with 0.2% (v/v) bromophenol blue from blue into yellow or greenish-yellow indicating pH drop through the release of organic acid into the medium by SSB inoculant [40].

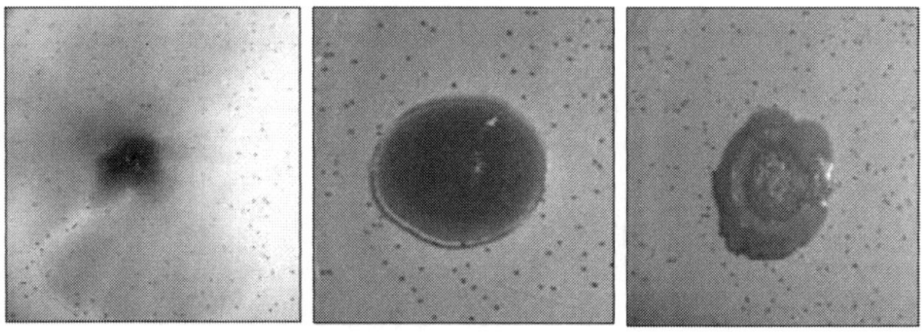

a b c

Figure 2. The growth of *Burkholderia cenocepacia* (a); *Aeromonas punctata* (b); and *Burkholderia vietnamiensis* (c) in solid Luria Bertani medium amended with 0.2% (v/v) Bromophenol Blue. The yellow zone indicates the ability of the isolates to generate organic acids.

According to Vandevivere [42], *Bacillus mucilaginosus* increases the dissolution rate of silicate and alumino-silicate minerals and releases the K^+ and SiO_2 from the crystal lattice primarily by generating organic acids. However, this hypothesis is controversial and *B. mucilaginosus* is also thought to accelerate the dissolution of a variety of silicates by the production of extracellular polysaccharides (EPS). The action of microorganism and chemical reactions in the soil make the silicon available to the plants. The primary and secondary silicate minerals in rocks are mineralized through microbial metabolism and chemical reactions, where the silica is converted and made available to plants. Bacterial metabolic activity in relation to micronutrient fixation in plants is the predominant role in these PGPR organisms like iron bacteria, sulphur bacteria, silicon mobilizing bacteria [43]. Bacteria when applied to root, cause an alteration in the rhizosphere, where its population persist and has a continuously increases in the ecosystem.

Silicon accumulation in plants is controlled by the ability of roots to take up Si. Deposition of silicon in root tissues, plants absorb Si as silicic acid [$Si(OH)_4$], an uncharged monomeric molecule, when the solution pH is below 9 [27]. Plants differ greatly in their ability to accumulate Si, ranging from 0.1% to 10.0% Si (dry weight) [28]. Factors such as soil pH, temperature, water conditions, the presence of cations, and organic compounds in solution influence the formation of soluble silicic acid and thus affect silicon accumulation in plants [44]. At the pH levels of most agricultural soils, H_4SiO_4 concentration in soil solution ranges from 0.1 to 0.6 mM [45]. Absorption of H_4SiO_4 takes place at the lateral roots via active, passive, and rejective mechanisms [46]. It is believed that in high Si accumulators, the amount of H_4SiO_4 taken up by active mechanism is greater than concentrations taken by mass flow because of the high density of Si transporters in roots and shoots facilitating H_4SiO_4 movement across root cell membranes [47]. The H_4SiO_4 absorbed by root's cells is deposited into leaf epidermal cells. The amount of literature documenting the benefits of Si on plants is vast and primarily highlights the value of Si fertilization in maintaining plant productivity under stressed conditions [48]. The established Si-induced mechanisms to improve plant resistance to biotic and abiotic stresses take place in the soil, the root system, and inside the plant [49]. Our research has shown that bio-silica application on oil palm seedling can enhance root growth under drought conditions. It has been demonstrated that deposition bio-silica occurs in endodermis part of roots oil palm seedling (**Figure 3**).

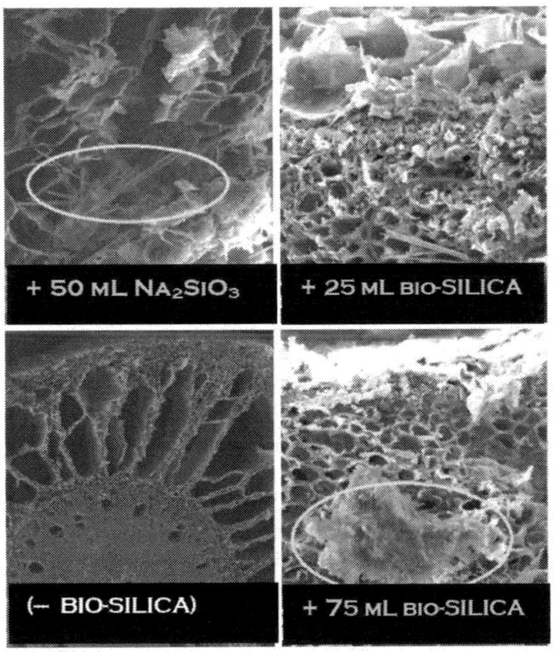

Figure 3. Scanning electron microscopy deposition of Si in root tissue of oil palm seedling by application bio-silica.

In drought-stressed sorghum [50] observed a significantly lower shoot/root ratio and higher root dry mass accumulation in silicon-applied plants compared with plants not treated with silicon, indicating that silicon facilitates root growth during drought. Ahmed *et al* [51] suggested that silicon application is mainly beneficial to the growth of sorghum root, allocating more matter to the plant root system grown hydroponically. The simulative effect of silicon on root growth may be due to enhanced root elongation as a consequence of enhanced cell wall extensibility in the growth zone, as observed in sorghum [52]. However, the beneficial effects of silicon on root growth under drought were not observed in some plants such as wheat, cucumber, and sunflower [12]. These observed differences may be related to culture conditions and plant species/cultivars. Nutrient uptake is related to root surface area and length. An increase in the surface area provides more exposed sites for uptake of diffusible ions. Silicon-mediated enhancement of root growth may, therefore, stimulate nutrient absorption and increase drought tolerance. In some studies, although silicon did not stimulate root growth under drought, silicon application, in fact, increased water uptake [53], thereby contributing to stimulation of nutrient uptake. The increased water uptake upon silicon addition under drought was due to the improved hydraulic conductance of roots [54] and root activity [55]. These studies suggest that silicon application may improve plant growth under drought by balancing nutrient uptake. Further investigations will be needed to determine how silicon regulates water uptake by roots and affects root anatomical characteristics to better understand the mechanisms of silicon promoted plant growth. Nutrient uptake is related to root surface area and length. An increase in the surface area provides more exposed sites for uptake of diffusible ions.

5. Conclusions

Due to several positive effects on the alleviation of different forms of biotic as well as abiotic stresses, silicon (Si) has been a focus of plant biology and agronomy research in recent decades. Silicon has numerous functions on plant physiology, and its most significant effects are focused on the cell wall. The presence of silicon in the cell wall increases their strength, resistance to salinity, drought tolerance, and photosynthetic activity. The beneficial effects of silicon are mostly on abiotic stresses such as drought. Oil palm seedlings treated with bio-silica grown in drought conditions displayed higher stomatal conductance than drought-stressed plants without bio-silica application. Silicon, especially refer to bio-silica can enhance root resistance to drought and promotes root growth. The other important role of bio-silica in reducing the adverse effects of stress may be by improving soil conditions. Therefore, bio-silica could be used as a growth regulator to improve plant growth and resistance to stress conditions. There is a need for applied research to evaluate water use efficiency and drought tolerance on more crops and areas.

6. References

[1] Yang S, Vanderbeld B, Wan J and Huang Y 2010 *Mol. Plant* **3** 469
[2] Wei L, Jia L, Hu X and Zhao F 1997 *Agric. Res. in the Arid Areas* **15** 66
[3] Kusaka M, Ohta M and Fujimura T 2005 *Physiol. Plant.* **125** 474
[4] Shao H B, Chu L Y, Shao M A, Jaleel C A and Hongmei M 2008 *Comptes Rendus—Biologies* **331** 433
[5] Jaleel C A, Manivannan P and Wahid A 2009 *Int. J. Agric. Biol.* **11** 100
[6] Farooq M, Wahid A, Kobayashi N, Fujita D and Basra S M A 2009 *Agric. Sust. Dev.* **29** 185
[7] Taiz L and Zeiger E 2006 *Plant Physiol.* (Sinauer, Sunderland, Mass, USA, 4th ed.)
[8] Garc´ıa-Plazaola J I and Becerril J M 2000 *Trees* **14** 485
[9] Ma J F and Yamaji N 2008 *Cell. Molecular Life Sci.* **65** 3049
[10] Siddiqui M H and Al-Whaibi M H 2014 *Saudi J. Biol. Sci.* **21** 13
[11] Liang Y, Sun W, Zhu Y G and Christie P 2007 *Environ. Pollution* **147** 422
[12] Pei Z F, Ming D F, Liu D, Wan G L, Geng X X, Gong H J and Zhou W J 2010 *J. Plant Growth Regulation* **29** 106
[13] Vasanthi N, Saleena L M and Raj S A 2012 *World Appl. Sci. J.* **17** 1425
[14] Balakhnina T and Borkowska A 2013 *Int. Agrophys.* **27** 225
[15] Karmollachaab A, Bakhshandeh A, Gharineh MH, Moradi Telavat MR and Fathi G 2013 *Internat. J. Agron. Plant Prod.* **4** 30

IOP Conf. Series: Earth and Environmental Science **183** (2018) 012014 doi:10.1088/1755-1315/183/1/012014

[16] Sacala E 2009 *J. Elementol.* **1** 619
[17] Bocharnikova E A and Benes S 2011 *Proc. Int. Conf. on Silicon in Agriculture* (Beijing: China)
[18] Son MS, Song JY, Lim MY, Sivanesan I and Jeong BR 2011 *Proc. Int. Conf. on Silicon in Agriculture* (Beijing: China) p1
[19] Mauad M, Crusciol C A C, Nascente A S, Filho H G and Lima G P P 2016 *Rev. Ciênc. Agron.* **47** 532
[20] Hajiboland R, Cherghvareh L and Dashtebani F 2017 *J. Plant Process Function* **5** 1
[21] Currie HA and Perry CC 2007 *Ann. Bot.* **100** 1383
[22] Heckman J 2013 *Better Crops* **97** 14
[23] Adrees M, Ali S, Rizwan M, Zia-Ur-Rehman M, Ibrahim M, Abbas F, Farid M, Qayyum MF and Irshad M K 2015 *Ecotoxicol. Environ. Saf.* **119** 186
[24] Ng L C, Anuar S N A, Jong J W and Elham M S H 2016 *Asian J. Plant Sci.* **15** 92
[25] Sakr N 2016 *J. Plant Protec. Res.* **56** 331
[26] Khalifa M M A, Nashwa A H, Fetyan, Magid M S A and El-Sheery N I 2017 *J. Agric. Res.* **6** 376
[27] Epstein E 1999 *Ann. Rev. Plant Physiol. Plant Mol. Biol.* **50** 641
[28] Richmond K E and Sussman M 2003 *Curr. Opin. Plant Biol.* **6** 268
[29] Gauger T, Byrne J M, Konhauser K O, Obst M, Crowe S and Kappler A 2016 *Earth and Planetary Sci. Lett.* **443** 81
[30] Sommer M, Kaczorek D, Kuzyakov Y and Breuer T 2006 *J. Plant Nutr. Soil Sci.* **169** 310
[31] Guntzer F, Keller C, Poulton P R, McGrath S P and Meunier J D 2012 *Plant Soil* **352** 173
[32] Rizwan M, Ali S, Ibrahim M, Farid M, Adrees M, Bharwana SA, Rehman M Z, Qayyum M F and Abbas F 2015 *Environ Sci. Pollut. Res.* **22** 15416
[33] Peera S K P G, Balasubramaniam P and Mahendran P P 2016 *J. Appl. Nat. Sci.* **8** 55
[34] Sahebi M, Hanafi M M, Nor Akmar A S, Rafii M Y, Azizi P, Tengoua F F, Azwa J N M and Shabanimofrad M 2015 *BioMed. Res. Int.* 1
[35] Luyckx M, Hausman J F, Lutts S and Guerriero G 2017 *Plants* **37** 1
[36] Kramer P J and Boyer J S 1995 *Water relations of plants and soils* (San Diego: Academic Press)
[37] Dodd I C 2003 *J. Plant Growth Regulation* **22** 32
[38] Muralikannan N 1996 *Thesis* (Coimbatore: Tamilnadu Agricultural University) p 125
[39] Chandramani P, Rajendran R, Muthiah C and Chinnaiah C 2010 *J. Biopesticides* **3** 423
[40] Santi L P and Goenadi D H 2017 *Menara Perkebunan* **85** 1
[41] Sheng X F, Zhao F, He LY, Qiu G and Chen L 2008 *Can J. Microbiol.* **54** 1064
[42] Vandevivere P, Welch S A, Ullman W J and Kirchman D L 1994. *Microb. Ecol.* **27** 241
[43] Gholami A, Shahsavani S and Nezarat S 2009 *World Acad of Sci Enginer and Tech* **49** 19
[44] Liu WG, Wang LQ and Bai YH 2003 *Acta Bot Boreali Occidentalia Sin* **23** 2248
[45] Knight CTG and Kinrade SD 2001 *Silicon in Agriculture* (Studies in Plant Science vol 8) ed Datnoff LE, Snyder GH, and Korndörfer GH (Amsterdam: The Netherlands Elsevier) p 57
[46] Cornelis JT, Delvauz B, Georg RB, Lucas Y, Ranger J and Opfergelt S 2011 *Biogeosciences* **8** 89
[47] Mitani N and Ma JF 2005 *J. Exp. Bot* **56** 1255
[48] Li QF, Ma CC and Shang QL 2007 *The journal of applied ecology* **18** 531
[49] Brenda ST, Babu T and Datnoff LE 2016 *Soil Science* **181** 1
[50] Hattori T, Inanaga S, Araki H, An P, Morita S, Luxová M and Lux A. 2005 *Physiol Plant* **123** 459
[51] Ahmed M, Hassen FU, Qadeer U and Aslam MA 2011 *Afr J Agric Res* **6** 594
[52] Hattori T, Inanaga S, Tanimoto E, Lux A, Luxova M and Sugimoto Y 2003 *Plant Cell Physiol* **44** 743
[53] Sonobe K, Hattori T, An P, Tsuji W, Eneji AE, Kobayashi S, Kawamura Y, Tanaka K and Inanaga S 2011 *J Plant Nutr* **34** 71

IOP Conf. Series: Earth and Environmental Science **183** (2018) 012014 doi:10.1088/1755-1315/183/1/012014

[54] Hattori T, Sonobe K, Inanaga S, An P and Morita S 2008 *J Plant Nutr* **31** 1046
[55] Chen W, Yao XQ, Cai KZ and Chen J 2011 *Biol Trace Elem Res* **142** 67

IOP Conf. Series: Earth and Environmental Science **183** (2018) 012015 doi:10.1088/1755-1315/183/1/012015

"CIRAGI"- The microbial fermentation starter for developing excellent coffee flavour

Tri Panji[1*], Priyono[1], Suharyanto[1] and A Zainuddin[2]

[1]Indonesian Research Institute for Biotechnology and Bioindustry, Jalan Taman Kencana No.1, Bogor 16128, Indonesia
[2]Department of Chemistry, Faculty of Mathematics and Natural Sciences, Padjadjaran University, Bandung, West Java, Indonesia

*E-mail: tripanji@iribb.org

Abstract. Most of the green bean coffees produced by smallholders are not properly fermented, while green bean coffees from large plantations undergo a spontaneous fermentation involving present microbial activity in the processing environment. The process of coffee fermentation itself involves not less than 144 endogenous microbial species originating from coffee fruits or from the surrounding environment. This condition is the source of inconsistency and low cup quality of coffee beans. The inoculation of selected microbial fermentation starter can transform microbial ecosystem. The transformation leads to beneficial microbes dominating and actively inducing flavour precursors into excellent aroma during roasting. The Indonesian Research Institute for Biotechnology and Bioindustry (IRIBB) has developed the CIRAGI (abbreviation of CItaRAsa tingGI meaning "a high flavour" in Indonesian), a fermentation starter for coffee beans using selected *Lactobacillus* sp. as the active ingredient. High cup quality coffee can consistently result from well-controlled fermentation process of the coffee bean using CIRAGI. The fermentation of both Robusta and Arabica coffee beans using CIRAGI has been demonstrated at laboratory and field trials. The cup tests of both Arabica and Robusta coffee have resulted in improved quality of the coffee flavour, increase of cup test scores based on the SCAA parameter, and elimination of off-flavour.

1. Introduction

Spontaneous fermentation of coffee beans using natural microbes is difficult to produce delicious taste precursors due to abundant natural microbes (about 144 strains) of different types depending on the processing environments [1, 2].

The efforts to improve the quality of coffee bean taste has been done, among others, through (1) application of yeast isolated from coffee fermentation [3], (2) the process of making artificial civet coffee, which mimics the process that occurs in the civet stomach [4] and civet coffee probiotics [5]. The Agustinisari, *et al.* used bioreactor to imitate the fermentation inside civet (*Paradoxurus hermaphroditus*) digestive system was being processed to obtain Indonesian Patent [4]. The fermentation starter used in this

IOP Conf. Series: Earth and Environmental Science **183** (2018) 012015 doi:10.1088/1755-1315/183/1/012015

research was bacterial isolates from civet digestive system liquid. This coffee making process was difficult to be applied in farmer's level due to the high price of the fermentor, also the limitation of fermentor capacity. The fermentation using microbial enzymes was considered expensive because of the microbial enzyme isolation cost. Moreover, the addition of crude enzyme could reduce the cup test score [6].

Most of the microorganisms that involved in the spontaneous fermentation of coffee were from coffee fruit or from the surrounding environment, i.e. fruit skin, mucus, and the shell of coffee bean. Besides that, the microbes may locate around the coffee processing environment site, i.e. in the fermentation box, the fermentation sacks, also the contamination from workers hand that involved in the fermentation process. Therefore, this spontaneous fermentation will produce various coffee flavours; depend on the type of indigenous microbes, the preparation of the fermented beans, the fermentation equipment, and the fermentation environment [7].

With adequate fermentation using important microbes from consistent inoculum, the fermentation process will run and the precursor's flavour of coffee will be formed. This process will produce a tasty and consistent coffee flavour during coffee bean roasting. Indonesian Research Institute for Biotechnology and Bioindustry (IRIBB) has a coffee bean fermentation starter named "Ciragi" with an active ingredient of superior *Lactobacillus* sp. that proved significantly increase the flavour of Robusta and Arabica coffees. The coffee bean fermentation starter "Ciragi" has been registered to Indonesian patent (Registration Number P-00201407291) and has obtained the trademark registration number (No. D002014052980).

This study presents the performance of Ciragi fermentation starter in improving the taste of Robusta and Arabica coffees from several coffee-producing regions in Indonesia compared to spontaneous fermentation result. The performance of the fermentation starter was measured based on the increase of coffee cup score that fermented with Ciragi compared to control without the addition of Ciragi.

2. Materials and Methods

The materials used in this study were Arabica and Robusta coffee bean from several regions in Indonesia from Sigararutang, AS2K, S795, and mixed varieties, which have been ripening optimally, and Ciragi fermentation starter with active ingredients *Lactobacillus* sp. Each coffee cherry was peeled using a pulper machine. The study was conducted with a minimum 100 kg of peeled-off coffee (pulped coffee cherries) for Ciragi treatment and control with two replications.

Fermentation using "Ciragi" was done by adding 0.5-1% (w/w) "Ciragi" starter to the peeled-off coffee (0.5-1 kg Ciragi per 100 kg peeled-off coffee) and then mixed evenly. Fermentations were done in plastic bags for 12-36 hours. Fermentation without "Ciragi" as control was done according to Standard Operational Procedure (SOP) applied in government own estate with 36 hours of fermentation time. After fermentation, each wet HS coffees were washed and dried under the sun until the moisture content reached 15%. Each HS coffee was peeled into green bean and re-dried under the sun until the water content reaches 12%.

As much as 250 g of green bean coffee samples, both for the treatment (fermented with Ciragi) and control (fermented without Ciragi), were sent to Indonesian Coffee and Cocoa Research Institute (ICCRI) for a flavour test based on SCAA cupping-protocols (http://www.scaa.org.). Several separated samples of green beans that fermented using "Ciragi" and control without Ciragi were sent to Laboratory in Netherland for cup test score based on Qmax and flavour notes.

3. Results and Discussion

The experiments of Ciragi fermentation have been done in several coffee estates in several provinces in Indonesia, namely West Java (Pangalengan and Ciwidey), Central Aceh (Bener Meriah and Takengon Regency), East Java (Jember and Banyuwangi), West Sumatera (Solok), East Nusa Tenggara

IOP Conf. Series: Earth and Environmental Science **183** (2018) 012015 doi:10.1088/1755-1315/183/1/012015

(Manggarai), South Sulawesi (Tana Toraja) and Central Java (Banaran). The type of coffee that used for this experiment depended on geographic condition. Some provinces only produce Arabica or Robusta coffee, but the other regions may produce both types of coffee.

3.1. Effect of fermentation starter against fermentation time

The preliminary study showed that fermentation using *Lactobacillus* formula could shorten the fermentation time from 36h to 12h (data not shown). The shorten fermentation time has several advantages, i.e. reduce the need of fermentation facilities, especially fermentation site and labour during the peak season of the coffee harvest period. For this reason, fermentation using Ciragi can be carried out for 12 hours. While the fermentation without Ciragi, the optimum fermentation time was achieved after 36 hours using a standard operational procedure that implemented in the government own estate.

3.2. Cup test score data from ICCRI

Table 1 showed that the application of "Ciragi" starter on coffee fermentation proved to significantly increase the cup score as well as the specialty level of coffee, both Arabica and Robusta coffees. These results are in line with the data in **Table 2** that show the cup test results from Netherland laboratory.

Table 1. Cup test score and flavour note of several coffees from coffee-producing centres in Indonesia with and without Ciragi fermentation starter

No.	Sample origin	Type of coffee	Cup test score and flavour note without "Ciragi"	Cup test score and flavour note with "Ciragi"
1.	Ciwidey, West Java	Arabica	82.75; spicy, herbal, astringent	86.25; sweetcorn, fragrance, lemony
2.	Bener Meriah 1, Aceh	Arabica	81.0; herbal, astringent, greenish, bitter after taste	85.0; flowery, sweet, sweet corn, caramelly, bright acidity
3.	Bener Meriah 2, Aceh	Arabica	80.75; fruity aroma, chocolaty	84.08, chocolaty, caramelly, clean
4.	Manggarai, East Nusa Tenggara	Robusta	79.38; dark chocolaty, rather rubbery, rather chemical, very astringent	84.50; chocolaty, acidity, excellent, buttery, fresh. On second experimental trial get cup score 89.30; Champion no 1 Banyuwangi Nusantara Coffee Festival
5.	Manggarai, East Nusa Tenggara	Arabica	82.75; spicy, herbal, astringent	86.25; sweet corn, fragrance, lemony. On second experimental trial get cup score 87.52; Champion no 1 Banyuwangi Nusantara Coffee Festival
6.	Tana Toraja, South Sulawesi	Arabica	83.50; spicy, herbal	85.38; spicy, nutty
7.	West Java	Arabica	83.50; vanilla, astringent, heavy body	84.50; caramelly, balance
8.	Bogor, West Java	Arabica	80.00; spicy	85.50; chocolaty
9.	Banaran, Central Java	Robusta	79.50 woody, caramelly	84.50 caramelly, spicy, vanilla

IOP Conf. Series: Earth and Environmental Science 183 (2018) 012015 doi:10.1088/1755-1315/183/1/012015

The latest data collected from applications of "Ciragi" fermentation starter were shown in the cup test results of Arabica Toraja coffee, Robusta Banaran coffee from government own estate, and farmer's coffee from Bogor region. In three samples observed, application of fermentation starter proved to increase the cup test score. The use of fermentation starter was also able to reduce astringency, herbal, and spicy-off flavours which are classified as coffee flavour defects.

3.3. Cup test score data from Netherland Laboratory

Similar data were obtained from Netherland laboratory (**Table 2**). Qmax from coffee bean fermented using Ciragi was higher compared to control without Ciragi. Some good flavour notes also appear from coffee fermented using Ciragi.

Table 2. Overall results of cup test score and flavour note from the Netherland laboratory for several

No.	Variety (field)	Control	Using fermentation Starter (Ciragi)
1.	ANDUNGSARI Variety, Sindangkerta field,	(No data)	Good smell, Good taste, Good body, Good acidity, Well balanced Fraction sweet/ fruity Good aftertaste. **Q mark: 8.0**
2.	ANDUNGSARI Variety, Pangalengan field,	Fraction common flat smell, Good body, fraction common/flat a little less sweet/fruity. Fairly neutral. **Q mark: 7,75**	Good smell, Sweet ferment, Good body, good acidity, green fruity aromatic but fraction ferment **Q mark: 8,25**
3.	SIGARARUTANG Variety, Pangalengan Field,	Good smell, Good body, Good aroma, good fruity acidity, fraction flat **Q mark: 8.0**	Good smell, Good body/aromatic Fraction green, Good acidity, Rather sweet **Q mark: 8,2**
4.	SIGARARUTANG Variety, Sindangkerta field,	Good aromatic smell, Good body, good acidity, Some sweet fruity acidity, Good balance Nice aftertaste. **Q mark: 8,4**	Good smell, Good body, good acidity, good sweet/ fruity aromatic with a very good aftertaste **Q mark: 8,5**
5.	Mix Variety, Riunggunung field,	Good smell, Good body, Somewhat sweet/fruity, Good acidity, Well balanced with a nice aromatic aftertaste **Q mark: 8,4**	Good smell, Very good, Green sweet/fruity, Good balance aromatic, Good aftertaste. **Q mark: 8,5**

Data in **Table 1** and **2** showed that *Lactobacillus* sp. as a fermenter was proven capable of increasing cup score of coffee bean. The improvement of coffee flavour using "Ciragi" fermentation starter was better than using physical treatment as described by Yusianto & Widyotomo (2013) [8]. Improving fermentation was the important way in increasing the quality of coffee bean [9]. Furthermore, the use of *Lactobacillus* sp. in coffee fermentation can reduce contamination by another microorganism, especially in suppressing the contamination of *Aspergillus ochraceus* that produce ochratoxin [10].

4. Conclusions

Coffee bean fermentation starter "Ciragi" with the active ingredient of selected *Lactobacillus* sp. was proven consistently capable in improvement cup score of Robusta as well as Arabica coffee in laboratory and field trials. The improvement of cup score and the elimination of off-flavour may result from well-controlled fermentation process of the coffee bean using these selected bacteria. Fermentation using selected bacteria has several advantages, not only in improving the formation of flavour precursor but also shorten the fermentation time from 36 to 12 hours.

5. Acknowledgements

Authors would like to thank Indonesian Coffee and Cocoa Research Institute (ICCRI) for flavour test of coffee samples and to Government own estate PTPN 9 for giving coffee cherries sample.

6. References

[1] Gaime-Perraud I, S Roussos and D Martinez-Carrera 1993 *Micol Neotrop Apl* **6** 95
[2] Rodarte M P, Dias D R, Vilela D M and Schwan R F 2011 *Maringá* **33** 457
[3] Pereira G V M, Soccol V T, Pandey A, Medeiros A B P, Lara J M R A, Gollo A L and Soccol C R 2014 *Intern. J. Food Microb.* **188** 60
[4] Agustinisari C, Nugraha S, Rahayu E, Tjahjohutomo R and Hadipernata M 2012 The process of making artificial civet coffee. Proposed Indonesian Patent. No: P00201200961.http://paten-indonesia.dgip.go.id/index.php/ web/search_result
[5] Guntoro, S 2010 Process of producing probiotic civet coffee. Paten-Indonesia. Proposal No: P00201000346. http://paten-indonesia.dgip.go.id/index.php/web /search_result/10
[6] Pushpa S M and Naidu M M 2011 *European Journal of Applied Sciences* **3** 130
[7] Towaha J, Purwanto E H and Aunillah A 2010 The role of processing on coffee flavor establishment. *Anthology of Coffee Plantation Technology for Smallholder Plantation*: 157p
[8] Yusianto and Widyotomo S 2013 *Pelita Perkebunan* **29** 220
[9] Widyotomo S, Purwadaria H K and Ismayadi C 2012 Increasing the quality and value added of coffee through the development of fermentation and decaffeination process. *InSINas Proceeding* 2012, 135p
[10] Godwin A, Sarah M and Lifa J 2010 *Int. J. Postharv.t Techno. and Innov.* **2** 41

The effect of biostimulant in root and population of phosphate solubilizing bacteria: a study case in upland rice

Agustin Sri Mulyatni[1*], R H Praptana[2] and D Santoso[1]

[1]Indonesian Research Institute for Biotechnology and Bioindustry, Jalan Taman Kencana No.1, Bogor 16128, Indonesia
[2]Indonesian Agency for Agriculture Research and Development, Jalan Merdeka No.47, Bogor, Indonesia

*Email: agustinmulyatni@gmail.com

Abstract. The productivity of upland rice is still low as compared to that of irrigated rice. A cultivation effort needs to be done to increase productivity. One effort can be made by providing additional biostimulants to promote growth and beneficial bacteria. The study was conducted to determine the effect of some biostimulants on the root and population of phosphate solubilizing bacteria in upland rice (Situ Patenggang varieties). The tested treatment was a combination of biostimulants derived from seaweed, humic acid, and organic compounds. Each treatment was repeated 5 times. Parameters observed were the length and weight root, pH analysis, phosphate and C-organic levels in the soil and population of phosphate solubilizing bacteria. The results showed that the combination of biostimulant gave significant effect on plant growth parameters. The combination of a humic acid with seaweed extract gave the highest length and weight root. Humic acid also has a significant effect on the total population of phosphate solubilizing bacteria. The total population of phosphate solubilizing increased up to 150,000 cfu/mg.

1. Introduction

Rice is the main staple food source in Indonesia. The needs of rice increase as the population grows. Upland rice cultivation is one of the solutions to increase rice productivity nationally. This is due to the increasing difficulty in finding wetland [1, 2]. The area of non-rice fields in Indonesia in 2014 in the form of the moor was 12 million ha [3]. The land is potential for cultivation of upland rice. Although the production level of upland rice is only 5% of national rice production regionally, its position is strategic enough to support food security. The average productivity of upland rice is 2.8 tons/ha, this result is still far below the average productivity of paddy rice that can reach 5-6 tons/ha [4]. Therefore, cultivation is needed to increase the productivity of upland rice. The use of biostimulant is one of the ways.

Biostimulant is a substance or microorganism given to plants in order to increase the absorption of nutrients so that the plants become more resistant to biotic and abiotic stress [5]. There are four categories of biostimulant which are humic acid, hormone-containing substance, amino acid, and microorganism.

The use of humic acid in rice, corn, potato, and wheat can increase the productivity of harvest [6-9]. Humic acid can increase the availability of phosphate, nitrogen, and micronutrient because of its ability to increase cation exchange capacity in the soil [10]. Biostimulant of material category contains hormones, for example, is seaweed extract. Application of seaweed extract on Almond can increase the number of shoots and plant biomass in low potassium soil condition [11]. According to Parrado [12], biostimulant derived from materials contains amino acids that can increase the growth and the number of tomato flowers. Similar benefits are also can be gained from biostimulant of microorganism.

The use of microbial inoculant in various food crops has been proved to increase crop productivity. Inoculation of microbes as PGPR (Plant Growth Promoting Rhizobacteria) combined with mycorrhiza can increase the yield and nutritional content in corn [13]. In rice with a salt high temperature of PGPR is able to increase the percentage of seed germination, root length, plant height and increase leaf chlorophyll content compared to control [14].

Increase in productivity of food crops may also use seaweed extract. This is because seaweed extract contains some phytohormones, polysaccharides, and micronutrient compounds that are very useful to push plant growth. Biostimulant application such as grass extract on tissue culture tobacco can induce flowering [15]. These results indicate that the content of phytohormones in seaweed extract is highly potential to induce more flowering. Similarly, application in Rice applied by seaweed of *Kappaphycus* sp and *Gracilaria* sp species can increase 41-34% of production [16]. In addition, the application of biostimulant from seaweed extract succeeds in increasing the number of Rice panicle in India [17].

Some of the results of this study indicate that the potential biostimulant to increase the productivity of upland rice is very large, even under the abiotic stress; it is also able to produce plants that resist abiotic stress. Therefore, this study aims to find an effective and efficient combination of biostimulant to increase the growth of upland rice and phosphate solvent bacteria population.

2. Materials and Methods

The research was conducted at the greenhouse of the Indonesian Research Institute for Biotechnology and Bioindustry from May to October 2016. There were 18 treatments with 5 replications. The treatment is as follows P2 (Soil + seaweed extract (ERL) 10 ppm 2x), P3 (Soil + ERL 10 ppm 3x), P4 (Ground + ERL) 0 ppm 2x), P1 (Soil + ERL 0 ppm 2x), P1 40 ppm 2x), P5 (Soil + ERL 40 ppm 3x), P6 (Soil + organic material + ERL 0 PPM 2x), P7 (Soil + organic material + ERL 0 PPM 3x), P8 (Soil + organic material + ERL 10 PPM 2x), P9 (Ground + organic material + ERL 10 PPM 3x), P10 (Soil + organic material + ERL 40 PPM 2x), P11 (Soil + organic material + ERL 40 PPM 3x), P12 (Soil + Humic Acid 0 PPM 2x), P13 (Soil + Humic Acid + ERL 0 PPM 3x), P14 (Soil + Humic Acid + ERL 10 PPM 2x), P15 (Soil + Humic Acid + ERL 10 PPM 3x), P16 (Soil + Humic Acid + ERL 40 PPM 2x), P17 (Soil + Humic acid + ERL 40 PPM 3x). Rice variety used was Situ Patenggang. Observation parameters were root length, root dry weight, pH, P, C-organic and total population of phosphate solvent bacteria.

3. Results and Discussion

Biostimulant treatment had significant effects on root length. The highest root length was found on humic acid treatment followed by organic matter. Root length increased and varied between 11-22% (**Figure 1**). Biostimulant treatment also increased root dry weight. The increase of root dry weight was due to a good plant growth so that the roots of the plant grew well. This was in accordance with a research conducted by Suwardi and Wijaya [17] where humic acid treatment can increase the length and dry weight of rice and corn roots. Anwar [18] also stated that humic acid increased the dry weight of wheat.

Biostimulant used affected total population of phosphate solubilizing bacteria. There was an increase in population up to 50,000-90,000 cfu/mg in the treatment of organic matter and an increase in the total population between 70,000-150,000 cfu/mg in humic acid treatment (**Figure 2**). Organic material and humic acid have a high carbon content that can be utilized by microorganisms as a source of energy [19] As a result, there were more population of phosphate solubilizing bacteria in the soil treated by biostimulant.

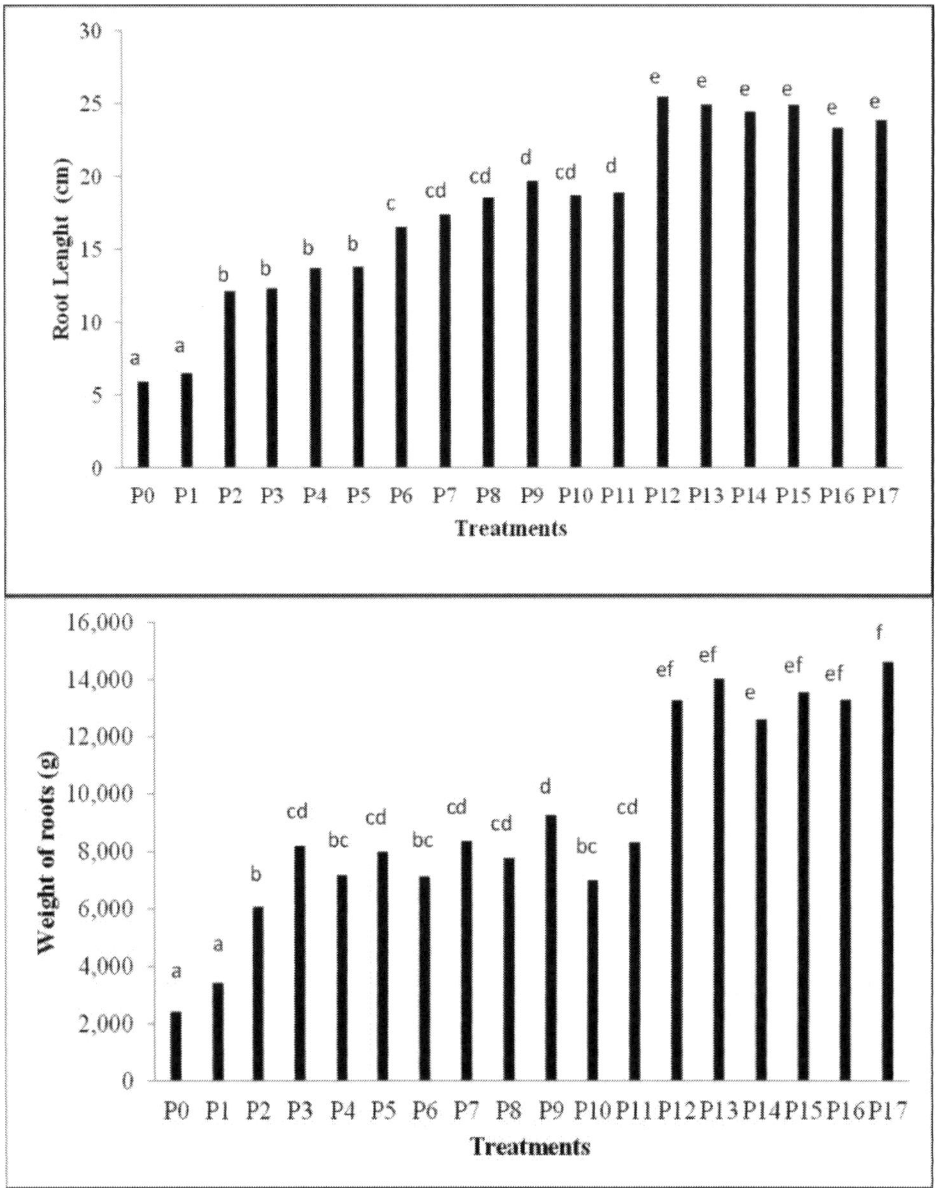

Figure 1. Length and weight of upland rice roots. Data are the mean value of 5 replicates for each treatment (P1-P18)

Provision of biostimulant increased the soil pH. The increase in pH was between 15-75% (**Table 1**). The increase in pH was due to the decomposition of organic matter causing the ligand exchange reaction between the anions to free OH - ions at the exchange site so it affects the increase of OH - number in the soil [20]. The increase in pH was also followed by an increase in C-organic and available-P. According to Winarso [20], the increase in pH can release P bound by Al so that P will be available by soil. The existence of biostimulant benefits, especially those derived from the humic acid

IOP Conf. Series: Earth and Environmental Science **183** (2018) 012016 doi:10.1088/1755-1315/183/1/012016

group is very useful to increase plant growth by increasing the available nutrients and beneficial bacteria especially phosphate solvent bacteria.

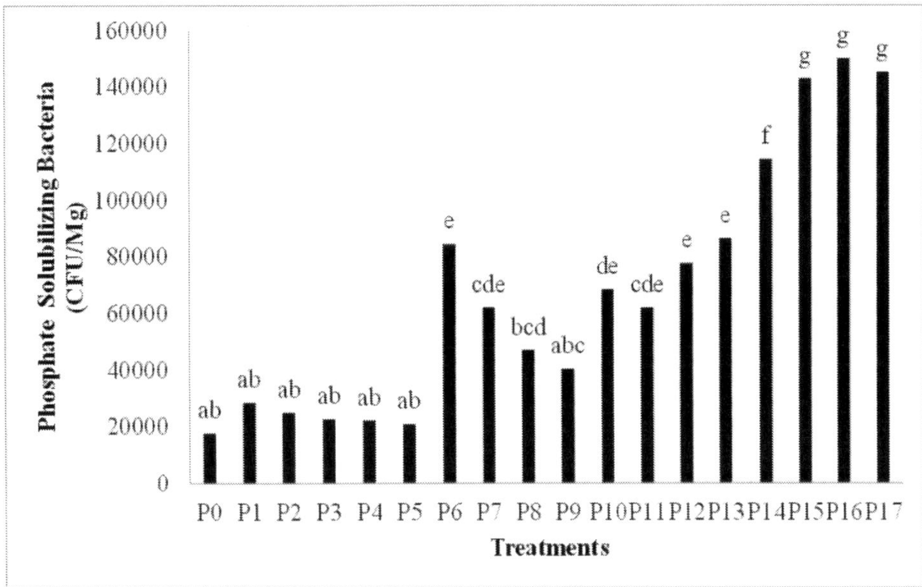

Figure 2. The population of phosphate solubilizing bacteria in upland rice. Data are the mean value of 5 replicates for each treatment (P1-P18)

Table 1. Levels of carbon, phosphate and pH in the soil of tested areas.

Treatments	C-organic (%)	pH	Available-P (%)
P1	1.29	6.02	6.7
P2	1.35	6.02	7.8
P3	1.54	6.05	9.8
P4	1.47	6.07	19.1
P5	1.52	6.05	14.9
P6	1.66	6.03	18.9
P7	1.95	6.17	22.2
P8	2.01	6.27	27.6
P9	2.23	6.18	45.9
P10	1.74	6.23	31.3
P11	2.12	6.15	29.9
P12	2.53	6.22	30.6
P13	3.67	6.76	43.2
P14	4.07	6.62	45.5
P15	3.72	6.68	44
P16	4.08	6.74	48.8
P17	4.01	6.56	57.4
P18	3.8	6.65	50.5

4. Conclusions

Biostimulant increased the length and dry weight of upland rice roots. In addition, humic acid biostimulant in particular increased phosphate solvent bacteria population.

5. Acknowledgements

The authors would like to thank Febriyanti Sitanggang for her technical assistance. The research was funded by the INSINas program of the Indonesian Ministry of Research, Technology and Higher Education.

6. References

[1] Herawati R, Purwoko B S and Dewi I S 2009 *J. Agron. Indonesia* **37** 2
[2] Samudin S and Adelina E 2016 Output Power and Quality of Some Local Gogo Rice Gototypes. *Proceedings of the National Seminar* on PPM IPB Results p 77
[3] BPS 2014 *Indonesian Statistics*. Badan Pusat Statistik Jakarta. Jakarta
[4] Nazirah L 2008 Responsiveness of upland rice varieties to the interval and the level of provision of water [*Thesis*] (Medan (ID):USU)
[5] du Jardin P 2015 *Sci. Hort.* **196** 3
[6] Garcia A C, Berbara R L L and Farias L P 2012 *J. Geocherm. Explor.* **136** 48
[7] Anjum S A, Wang L, Farooq M, Xue L and Ali S 2011 *J. Agro. Crop Sci.* **197** 409
[8] Selim E M, Shayma I S, Asaad F F and El Neklawy A S 2012 *J. Appl. Sci. Res.* **8** 531
[9] Yildrim 2007 *Acta Agric. Scand.* **57** 182
[10] Berbara R LL and García A C 2014 Humic substances and plant defense metabolism *Physiological Mechanisms and Adaptation Strategies in Plants Under Changing Enviornoment* vol 1 ed Ahmad P and Wani M R (New York (US): Springer pp 297-319
[11] Saa S, Olivos-Del Rio A, Castro S and Brown P H 2015 *Front. Plant. Sci.* **6** 87
[12] Parrado J, Bautista J and Romero E P 2008 *Bioresour. Technol.* **99** 2312
[13] Adesemoye A O, Torbert H A and Klopper J W 2010 *Appl. Soil. Ecol.* **46** 54
[14] Santoso D and Priyono 2014 Proses Produksi dan Formulasi Biostimulan dari Alga Coklat *Sargassum* sp. serta Penggunaannya untuk Pertumbuhan Tanaman. P-00201406718 (*Indonesian Patent*).
[15] Pramanick B, Brahmachari K, Ghoshi A and Zodape S T 2014 *Bangladesh J. Bot.* **43** 53
[16] Srivastava A, Bhatia G, Pant P and Srivastava P C 2010 *J. Environt. Protection* **1** 261
[17] Suwardi and H Wijaya 2013 *J. Ilmu Pertanian Indonesia* **8** 79
[18] Anwar S, Iqbal F, Khattak W A, Islam M, Babar-Iqbal and Khan S 2016 *EC Agric.* **3** 558
[19] Angin I, Turan M, Ketterings Q M and Cakici A 2008 *Water Soil Pollution* **188** 335
[20] Winarso S, Sulistyanto D and Handayanto E 2011 *J. Ecol. Nature Environ.* **3** 232

IOP Conf. Series: Earth and Environmental Science **183** (2018) 012017 doi:10.1088/1755-1315/183/1/012017

Plant biostimulant to improve crops productivity and planters profit

Djoko Santoso[1], A Gunawan[2], A Budiani[1], D A Sari[1] and Priyono[1*]

[1]Indonesian Research Institute for Biotechnology and Bioindustry, Jalan Taman Kencana No.1, Bogor 16128, Indonesia
[2]PT Maxima Agro Internasional, Jalan Teuku Umar 432, Sokaraja, Banyumas, Central Java, Indonesia

*E-mail: priyono@iribb.org

Abstract. Plant biostimulants are applied to meet the challenge for sustainability of agriculture production. IRIBB has developed organic-based biostimulant formulas (OBFs) and tested in the field of several crops. OBF-1 improved the vegetative growth and productivity of some annual food crops. Compared with the control plants, the productivity of OBF-1 treated rice, maize, potato and shallot increased by 25%, 31%, 30% and 23%, respectively. In addition, OBF-1 treated crops indicated more tolerant to pests and diseases. OBFs applications on perennial estate crops demonstrated that OBF-2 and OBF-1 combination increased tea yield around 48% and sugarcane yield up to 50%. Whereas foliar sprays of OBF-3 on oil palm increased the CPO yield by up to 32%. The increased CPO yields were contributed to the increase of total FFB (fresh fruit bunches) weight and oil extraction rate in the fruit mesocarp. OBF-3 is a formula enriched with oil-biosynthetic activators. The farmers and planters should gain additional profits from the increased crop yields in response to the OBF application. How much profits gained were depended also on the price of the commodities. In general, the additional profits were much higher than the OBFs and labour costs.

1. Introduction

Within 40 years, the world population was doubling, increased from 3 billion in 1959 to 6 billion by 1999. The US Census Bureau [1] estimated that world population will continue to grow even though at a slower rate. The world population is projected to grow from 6 billion in 1999 to 9 billion by 2042 [1]. The increasing number of the world population is contributed by both the increased number of new births and life expectancy [2]. To support the needs of world growing population for primary food and energy, the supply of agricultural products must be increased. With decreasing area of arable land for agriculture production, an intensification strategy is necessary. The use of plant biostimulant is considered an innovative approach to increase agriculture production in a sustainable and environmentally friendly way [3].

Plant biostimulants can be defined as substances or products that contain compounds, microorganisms or formulations stimulating natural process of plant growth. Of the developing definitions, there are five categories of biostimulants: microbial inoculants, humic acids, fulvic acids, protein hydrolysates and amino acids, and seaweed extracts [4, 5]. The Indonesian Research Institute for Biotechnology and Bioindustry (IRIBB) has developed some formulations of plant biostimulant with primary components are extracts of natural organic materials from local sources. The extracts and formula were prepared in such unique ways in both compositions and bioactivities to plant growth and development [6, 7]. The organic-based biostimulant formulas (OBFs) have been tested to stimulate the growth and development of some plants *in vitro*, in the greenhouse, and in the fields.

This paper reported some of the results of those tests. The test at the laboratory level was conducted using tobacco (*Nicotiana tabacum* L.) plantlets grown *in vitro*. The field trials were mainly to evaluate the effects of OBFs on productivities of some perennial estate crops and annual food crops such as rice, maize, potato, and shallot. The perennial estate crops used to test the efficacy of the OBFs were tea (*Camellia sinensis* L.) and oil palm (*Elaeis guineensis* Jacq.) in the plantations, and sugarcane (*Saccharum officinarum* L.) grown in polybags.

2. Materials and Methods

Organic extracts were prepared as described previously [6]. Chemicals for plant tissue culture media were bought from Sigma Chem. Co. or local sources. Biostimulant stock solutions of several formulations were produced and kindly provided by Biostimulant Lab Production of IRIBB. These stocks sent to the fields were prepared with concentrations 10,000 ppm (OBF-1 and OBF-3) and 80,000 ppm (OBF-2). To most crops in the fields, 250 to 1,000 times diluted biostimulant solutions were applied through foliar spraying. For mature oil palm trees taller than 5 meters, there were 2 alternative ways of application, trunk injection or root infusion.

2.1. In vitro *testing of organic extracts*

In vitro culture medium for tobacco, plantlet development was prepared using basic MS salt [8] with a half concentration as used for Lilium plant species [9]. The MS salt solution was added with 3% sucrose and 1% Gelrite. After the pH was adjusted to 5.7, sterilization was performed using an autoclave at 121 °C for 30 min. A certain amount (0.5 to 40 ppm) of filter-sterilized organic extracts were added and mixed into warm autoclaved media. After the media were solidified and cooled in some culture tubes, sterile tobacco plantlets were cultured into the tubes and maintained in a culture room. Plantlet development was observed every 2 to 7 days.

2.2. Foliar sprays

For foliar spray application, diluted OBF solution was filled into the tank of a power sprayer. The doses of the sprays were between 200 - 300 L/ha for all crops or 1.5 L per productive oil palm tree. The sprays were performed during the shady time, preferably early in the morning when the stomata were open.

2.3. Trunk injection

Application of the biostimulant through trunk injection used diluted biostimulant solution with smaller volumes but higher concentration, volume 60 - 100 mL per tree and dilution 200 - 500 times. Small holes with diameter 1 – 1.5 cm and depth 10 - 15 cm were made with 30-degree slope into the palm trunk. Using a funnel 30 - 50 mL diluted biostimulant was filled per hole.

2.4. Root infusion

In term of dosage, application of biostimulant through root infusion was more similar to trunk injection than to foliar spray. The volume could be 100 to 200 mL per spot and dilution 400 – 600 times. Roots to be infused were chosen the healthiest ones. The chosen roots were selected by digging carefully the soil at the spotted area approximately 50-cm from the outer side of the trunk. Two healthy roots at the opposite positions on the trunk were cut smoothly with a very sharp cutter. The cut root was inserted into a 500-mL plastic bag with 5 to 8 cm in diameter that contains 100 to 200 mL diluted biostimulant. After insertion of a straw, the open end of the bag was tied loosely together with the cut root and the straw, the construction was dumped loosely with soil and the tip of the straw was

IOP Conf. Series: Earth and Environmental Science **183** (2018) 012017 doi:10.1088/1755-1315/183/1/012017

several centimetres above the ground and covered with cotton. The function of the straw was to later normalize air pressure in the bag and refill the bag with biostimulant solution.

3. Results and Discussion

Bioactivities of organic extracts were tested *in vitro* using tobacco plantlets observed weekly until 10 weeks in culture. **Table 1** indicates that the two preparation of organic extracts were strong inducers of the early development of roots. The four treatments induced plantlet rooting to some extent were similar in one week of culture, where the control plantlets had no root yet. The Exalk-4 provided several different types of plant development at different lengths of culture. While the other three extracts were similar, since two weeks in culture stimulating only the vegetative growth and no generative development. In addition to vegetative development and growth, the Exalk-4 strongly stimulated generative development. Only within the two-week culture, Exalk-4 induced flower buds on the top of the tobacco plantlets. Whereas in normal culture condition (without extract), the tobacco plantlets were never flowering. The strength of florigenic properties of the Exalk-4 could be confirmed with a comparison to transgenic tobacco carrying flowering transgene. In transgenic tobacco plantlet ectopically expressing flowering gene *TcAP1*, the tobacco plantlet was flowering only after 3.5 months or about 14 weeks in culture [10]. These data indicate also that the Exalk tends to work better in a lower concentration. Whereas the Exair to some extent was more effective in a higher concentration.

Table 1. Growth and development responses of tobacco plantlets to the presence of extracts.

Treatments	1 week rooting	2 weeks flowering	7 weeks buds	10 weeks
Control	None	none	none	none
Exalk 4	N=23; L 4-7 mm	Flower bud	Apical flower bud wilted, many axillary buds	A flower bud formed on every axillary bud
Exalk 40	N=16; L 2-7 mm	none	none	none
Exair 4	N=5; L 1-15 mm	none	none	none
Exair 40	N=18; L 3-28 mm	none	none	none

Note N= number of developing roots per plantlet; L= length of roots.

Considering that the organic extracts contain some kinds of phytohormones, the root-inducing capability of the extracts was likely provided by the cytokinin content in the extracts. Research by de Vries *et al.* [11] reported that cytokinin induced promotion of root meristem in the fern. Flower induction involves a more complex network of molecular regulation including hormones. In apple trees (*Malus domestica*) flower formation involved cytokinin, abscisic acid and gibberellic acid pathways [12]. Gibberellic acid promoted the transition from vegetative development to the first inflorescence phase of reproductive development of *Arabidopsis* [13]. In the case of Exalk 4, strongly induced flower buds of tobacco plantlet could be likely synergistic effect of the composition of flower-promoting hormones in the extract cytokinin and gibberellic acid. Our previous analysis using LCMS indicated that the extract contains GA_3 (gibberellic acid-3) and cytokinin.

These *in vitro* data of the tobacco plantlet development were considered to be the key information useful in formulating OBFs. Type of extract and concentration in the treatments extrapolated to the composition of those in OBFs. There are usually certain conversion factors that can be used to transform composition and concentration from the laboratory level to the field trial level. In general, the dosage needed by the crops in the field is higher than that of the model plants in the laboratory.

IOP Conf. Series: Earth and Environmental Science **183** (2018) 012017 doi:10.1088/1755-1315/183/1/012017

3.1. OBFs field trials on annual food crops

Application of OBFs to improve the yield of some annual food crops such as rice, maize, potato and shallot was examined in several different locations in Java. These trials involved directly small farmers own the lands. The results were presented in **Table 2**. These data show that the increases in those crops productivity were varying from the lowest 23% in shallot to the highest 31% in maize. The most comprehensive data were collected from trials on rice crop. In addition to data on the efficacy, data of surveys were collected from farmers involved directly in this trials. One of the questions of the surveys was "after being experienced using OBF-1 to improve rice productivity, how much satisfied you (the farmers) on the efficacy of OBF-1?". From 129 farmers of the respondents, 62 farmers gave the highest level of satisfaction (score 5 of 5 choices available), 66 farmers provided score 4, and 1 respondent chose none [14]. These farmers satisfaction could be the reflection of the additional profit gained from the increased yield of the crops.

Table 2. Improved yield of annual food crops by biostimulant OBF-1.

Crops	Location in Java	The control yield (ton/ha)	Yield increase (%)
Rice	East, Central, West	7.05	25
Maize	West	9.20	31
Potato	Central, West	18.25	30
Shallot	Central	9.67	23

Additional profits possibly gained by the farmers from the increased productivity calculated using the selling price of the commodities at the local or neighbouring market. The selling price at the producer level was obtained from the online information service provided by the Indonesian Ministry of Agriculture and can be accessed at http://aplikasi.pertanian.go.id/smshargakab/qrylaprmsayu.asp. [15]. **Table 3** shows that additional profit by the OBFs application could provide 4- to almost 20-fold additional profits to the additional cost. The lowest additional profit was from OBFs on rice, but it was still more than 4-fold of the additional cost. The highest additional profit was gained from OBFs on potato, it was almost 20-fold of the additional cost.

Table 3. Increased profit of the annual food crops treated with biostimulant OBF-1.

Crops	Yield increase (kg/ha)	Selling price (IDR/kg)	Additional cost (IDR/ha)	Additional profit[1] (IDR/ha)
Rice	1,507	4,500	700,000	6,081,500
Corn	2,852	3,700	900,000	9,752,400
Potato	5,475	7,500	6,000,000	35,062,500
Shallot	2,224	12,000	2,000,000	24,689,200

[1]Additional cost of OBFs and labours varied from IDR 1,500,000 to 2,000,000

Efficacy of OBFs on three different types of perennial estate crops has been conducted on plantations and on polybags. The range of their productivity increases was comparable with those of the annual crops above (**Table 4**). The highest increase occurred in OBFs on sugarcane, which was up to 50%. Two conditions were applied on the sugarcanes. The plants were grown in more favourable conditions, controllable polybags and using two other types of plant biostimulants which were humic acid-based plant biostimulant and microbial inoculant of mycorrhiza. Whether the increased productivity of the sugarcane in the field would be similar, a field trial should be conducted with the same treatments. In fact, a field trial is being performed in East Java.

Table 4. Improved yield of estate crops by biostimulant OBF-1.

Crops	Location	The control yield (kg/ha)	Yield increase (%)
Tea	PTPN-8; PTPN-6	1,941	48
Oil palm	PTPN-6	6,805	29
Sugarcane*	IRIBB	7,800	50

*Grown in polybags

From the data in **Table 4** and the listed current selling price of the commodities, the additional profits were calculated. The additional profits gained by the planters are profits obtained from additional production multiplied by the selling price, then are subtracted by additional cost. **Table 5** indicates that the most profitable in number is the biostimulants on sugarcane, the lowest profit is OBF on oil palm. Nevertheless, the use of OBFs on oil palm is very attractive because the additional profit is still much higher than the additional cost, more than 3-fold of the cost. Besides, the potential market of the OBFs on oil palm plantations is the biggest. Owned by mostly big companies, the purchasing power of the plantations for OBFs should be attractive for doing business with.

Table 5. Increased profit (IDR) of estate crops by biostimulant OBF-1.

Commodity	Additional production (kg/ha)	Selling price (IDR/kg)	Additional cost (IDR)	Additional profit (IDR)
Tea	932	18,000	2,460,000	14,306,860
Oil palm	1,983	8,000	3,764,400	12,099,600
Sugarcane*	3,900	11,000	3,500,000	39,400,000

*Grown in polybags

OBFs have attractive in both scientific and business aspects. That mechanisms by which OBFs regulates plant development has not yet been revealed are interesting objects of further studies. These include more detail composition of the bioactive compounds in each specific extract of the organics used.

The existing formulations the OBF-1 to -3 proven capable of improving the productivity of some trial crops are profitably attractive. Therefore, mutually beneficial involvements of the industries are necessary for commercialization steps. There were indications that the OBFs improve the quality of the harvests needs to be further explored both scientifically and economically. Moreover, new formulations maybe developed with a specific function to a specific crop. Otherwise, new OBFs formulations and combinations that have more general functions to most crops can be alternative targets of development. All of these should be developed in response to farmers and planters needs to get beneficial impacts that supportive to sustainable developments.

4. Conclusions

Organics-based biostimulant formulations (OBFs) were proven to improve the productivity of some annual food crops and perennial estate crops. The yields of OBFs-treated rice, maize, potato and shallot in the fields increased around 25%, 31%, 30% and 23%, respectively. Likewise, OBFs showed good effects on improving the productivity of tea, oil palm and sugarcane by 48%, 30% and 50%. Based on the results, the farmers and planters should gain additional profits from the increased yields of the crops in response to the OBFs application. How much more profits gained were depended also on the selling prices of the crops. The additional profits were much higher than the OBFs and labour costs, could be 4- to 20-fold of the cost.

5. Acknowledgements

This research was partly funded by RISPRO Program of LPDP, Ministry of Finance and the Program of Industry Innovation of Ministry of Research, Technology and Higher Education, The Republic of Indonesia. The authors wish to thank the field trials team of technicians of IRIBB and agronomists of PT MAI, the industrial partner working on the marketing of plant biostimulants and related products.

6. References

[1] US Cencus Bereau 2017 World Population: 1950 - 2050. https://www.census.gov/population/international /data/worldpop/graph_population.php
[2] InfoDATIN 2014 Situasi dan Analisis Lanjut Usia. Kementerian Kesehatan Republik Indonesia, Jakarta Selatan. 8p. (*in Indonesian*)
[3] Povero G, Mejia J F, Tommaso D D, Piaggesi A and Warrior P 2016 *Frontiers in Plant Sci.b* **7** 435
[4] Sharma H S S, Fleming C, Selby C, Rao J R and Martin T 2013 *J. Appl. Phycol.* **26** 465
[5] Calvo P, Nelson L and Kloepper J W 2014 *Plant. Soil* **383** 3
[6] Santoso D and Priyono 2014 Proses Produksi dan Formulasi Biostimulan dari Alga Coklat *Sargassum* sp. serta Penggunaannya untuk Pertumbuhan Tanaman. P-00201406718. (*Indonesian Patent*)
[7] Santoso D, Priyono and Budiani A 2017 Formulasi Biostimulan Palmarin dan Penggunaannya untuk Peningkatan Rendemen Minyak dan Bobot Tandan Buah Segar (TBS) Kelapa Sawit. P-00201701786. (*Indonesian Patent*)
[8] Murashige T and Skoog P 1962 *Physiol. Plant.* **15** 473
[9] Nhut D T, Le B V, Tanaka M and Van K T T 2001 *Sci. Hort.* **87** 131
[10] Chaidamsari T, Samanhudi, Budiani A, Poerwanto R and Santoso D 2006 *Menara Perkebunan* **74** 1
[11] de Vries1 J, Fischer A M, Roettger M, Rommel S, Schluepmann H, Brautigam A, Carlsbecker A and Gould S B 2016 *New Phytologist* **209** 705
[12] Xing L, Zhang D, Li Y, Shen Y, Zhao C, Ma J, An N and Han M 2015 *Plant Cell Physiol.* **56** 2052
[13] Yamaguchi N, Winter C M, Wu M F, Kanno Y, Yamaguchi A, Seo M and Wagner D 2014 *Science (N.Y.)* **344** 638
[14] Santoso D, Priyono, Kresnawaty I and Mulyatni A 2016 Produksi dan Aplikasi Biostimulan untuk Meningkatkan Produktivitas dan Kualitas Panenan serta Pendapatan Petani Padi, Laporan Penelitian Tahun I, LPDP, Kementerian Keuangan, Juni 2016 (*in Indonesian*)
[15] Kementerian Pertanian 2017 Informasi Harga Komoditas Pertanian Kabupaten. *Laporan Harian* http://aplikasi.pertanian.go.id/smshargakab/qrylaphar.asp

IOP Conf. Series: Earth and Environmental Science **183** (2018) 012018 doi:10.1088/1755-1315/183/1/012018

Sucrose analysis in leaf tissue of two contrasted rubber tree clones under abiotic and biotic stresses

Irfan Martiansyah[1,2], R A Putranto[1*] and N Khumaida[2]

[1]Indonesian Research Institute for Biotechnology and Bioindustry, Jalan Taman Kencana No.1, Bogor 16128, Indonesia
[2]Plant Breeding and Biotechnology, Faculty of Agronomy and Horticulture, Bogor Agricultural University, Dramaga, Bogor 16680, Indonesia

*E-mail: rizaputranto@iribb.org

Abstract. The sucrose content in rubber leaf is assumed to differ between genotypes and environmental conditions. We studied the relation of sucrose-content and -concentration between high- and low- latex metabolism clones, such as PB 260 and PR 300, respectively under biotic- and abiotic-stresses (white root disease and tapping panel dryness). The experimental design was a complete randomized design of two treatments (by genotypes and by stress conditions) with five replicates. Sucrose analysis was performed using the anthrone colourimetric modified method at 627 nm of spectrophotometry. Statistical analysis used the Student Newman-Keuls test of ANOVA at 5% significant level. The different types of latex metabolism clones had not affected the value of sucrose-content and -concentration in leaf tissue. By contrast, the results showed that WRD-affected trees of PB 260 clone (24.45 mg g^{-1}) had significantly the highest sucrose content compared with healthy- (15.53 mg g^{-1}) and TPD-affected trees (13.42 mg g^{-1}). A similar trend was equally observed in TPD- (16.33 mg g^{-1}), healthy- (14.51 mg g^{-1}) and WRD-affected trees (14.52 mg g^{-1}) of PR 300 clone. The sucrose-content and -concentration in PB 260 clone at WRD-affected trees were the highest than the others.

1. Introduction

Natural rubber (*cis*-1,4-polyisoprene) is commercially used as a raw material of various industrial products. It is synthesized in the cytoplasm of highly differentiated cells of the cambium namely laticifers [1, 2]. Latex containing natural rubber particle is harvested by tapping the outer layer of bark truncating the laticifers [2, 3]. The latex flows out directly after the first tapping until the beginning of coagulation processes by lutoid activity inside the laticifer cells [3, 4]. In periodical tapping, latex regeneration and duration of latex flow are critical factors reflecting latex production in rubber plantation [5]. Latex regeneration takes at least 48 hours to convert sucrose into *cis*-1,4-polyisoprene in natural rubber biosynthesis [6, 7]. Latex regeneration between the two intercepts of tapping period is regulated by primary mechanisms associated with the sucrose availability, biochemical reaction providing energy, enzyme activity, water supply, nitrogen availability, and anti-senescence activity [6, 8-14]. A carbon and energy supply for latex regeneration is provided by sucrose derived from rubber tree leaves [15].

Sucrose is an early precursor for the biosynthesis of rubber particles [16]. The activity of latex vessel depends on the availability of sucrose inside laticifers [17-19]. The synthesis of isoprene molecule requires the sucrose to be broken down into glucose and fructose by the invertase enzyme [20]. These processes generate the biochemical energy to be used for producing rubber particle [18, 21, 22]. The synthesis process begins with the breakdown of sucrose into other molecules through more than twenty enzymatic reactions to generate the first isoprene molecule of isopentenyl diphosphate (IPP) [23, 24]. IPP is a direct intermediate precursor for latex biosynthesis. It is a major derivative of the MVA (mevalonate) pathway or MEP (2-C-methyl-D-erythritol-4-phosphate) as an alternative biosynthetic pathway [23]. Both of these latex biosynthetic pathways occur in laticifers and require sucrose as the precursors [24]. The entry of sucrose into laticifers and enzyme activity are two important steps in latex synthesis [25]. In addition, sucrose and potassium play a role in regulating the turgor pressure (0.9-1.5 MPa) in the laticifers and surrounding cells in the bark to maintain the latex flow [26, 27]. Meanwhile, sucrose in the cytoplasm is used for metabolite activities evaluation equally present in the laticifers [11, 22].

Laticifers is an active sink to accumulate sucrose from photosynthesis organ. The sucrose transport into the laticifers is actively carried out throughout the plasma membrane [11, 17, 22, 27, 28]. The physiological and molecular regulation of sucrose metabolism has been widely studied in latex association to natural rubber production in laticifers [6, 18-20, 29]. Sucrose loading into laticifer is one of the limiting factors of latex production in *Hevea brasiliensis* [9, 30]. High sucrose-content inside the laticifers indicates that this precursor is not utilized for latex biosynthesis causing later reduced latex production [6, 7]. The sucrose availability supposes to depend on the photosynthesis capability in leaves. However, studies of the sucrose metabolism in rubber tree leaves are limited. Logically, to regenerate latex, a high level of sucrose-content in photosynthesis organ must be maintained. The sucrose in rubber leaf is assumed to differ between genotypes and environmental conditions. Photosynthesis process is inhibited when rubber trees undergo stress conditions disturbing the translocation of sucrose into laticifers. This research aimed to study the relation of sucrose-content between high- and low-latex metabolism of rubber tree clones under abiotic and biotic stresses (tapping panel dryness-TPD and white root disease-WRD).

2. Materials and Methods

2.1. Plant material

Leaf samples were obtained from rubber tree clones PB 260 and PR 300 (**Figure 1**) grown at the field trial of Ciomas, Indonesian Research Institute for Biotechnology and Bioindustry, Bogor, Indonesia (Latitude: S 6° 36' 22.98"; Longitude: E 106° 45' 42.47"). PB 260 is a quick starter rubber tree clone as opposed to a slow starter PR 300. The samples were collected in three conditions such as healthy trees, tapping panel dryness (TPD)-affected trees at 33% of latex halt [25], white root disease (WRD)-affected trees carrying early visual attack symptoms. TPD as a physiological disease represents abiotic stress while WRD caused by *Rigidoporus lignosus* represents biotic stress. The collected samples were carried into the laboratory on dry ice and directly used for sucrose analysis.

2.2. Sucrose analysis

2.2.1. Anthrone reagent

The sucrose-concentration was measured using the modified anthrone colourimetric method [31]. This reagent was prepared by dissolving 0.1 g of anthrone (Merck, USA) in 100 mL of H_2SO_4. The acid was previously made by adding 100 mL of acid to 30 mL of water. The reagent was allowed to stand for 30-40 min with occasional shaking until complete clearance of the mixture. The reagent was freshly prepared each day and used within 12 hr.

Figure 1. Leaf samples of two contrasted rubber tree clones. (*Left*) A quick starter clone (PB 260). (*Right*) A slow starter clone (PR 300).

2.2.2. Sucrose standard
Sucrose stock solutions (1 mM) was prepared by dissolving 1.73 mg of sucrose powder (Caisson Labs, USA) in 125 μL trichloroacetic acid (TCA) 2.5%.

Table 1. Series of sucrose-concentration for the standard curve.

Solution	Series of sucrose-concentration (mM)						
	0	0.01	0.025	0.05	0.1	0.2	0.5
Sucrose stock solution (1 mM)	0	5	12.5	25	50	100	250
TCA 2.5% (μL)	500	495	487.5	475	450	400	250

2.2.3. Method
Briefly, 0.5 g leaf tissue samples were extracted in a 50 mL centrifuge tube, and 10 mL of 80% ethanol was added. The samples were heated in a water bath at 85 °C for 30 minutes followed by a centrifugation at 10,000 rpm for 15 minutes. The supernatant was collected. The extraction was repeated twice (10,000 rpm for 15 min each) and heated in the boiling water until it has condensed to 5 mL. Distilled water was added to a final volume of 50 mL. Fifteen μL of the solution was taken followed by a sequential addition of 135 μL of distilled water and trichloro-acetic acid 2.5 % (TCA) to a final volume of 500 μL. Finally, 3 mL anthrone reagent was added into the solution followed by incubation in boiling water for 15 min. The absorbance value at 627 nm was measured using spectrophotometer (Multiskan Go, ThermoScientific).

2.3. Data analysis and statistical
The sucrose-concentration (mM) for each of samples determined by the sucrose standard equation formula: **y** = 2.3852x + 0.0235 with 'x' is the absorbance data values. The sucrose-content value was calculated by the equation:

$$\text{Sucrose-content (mg g}^{-1}\text{): } \frac{(C \times V \times Mw) \times Fd}{W}$$

where:
C is sucrose-concentration obtained by referring to the standard curve (mM), V is the volume of solvent (L), Mw is sucrose molecular weight (342.3 g mol-1), W is the weight of the sample (g), Fd is the dilution factor (10x).

The mean values of sucrose-concentration obtained from five replicates of each treatment. Statistical analysis was performed using the Analysis of Variance (ANOVA) at 5% level followed by Newman-Keuls Student test (XLStat, Addinsoft, USA).

IOP Conf. Series: Earth and Environmental Science **183** (2018) 012018 doi:10.1088/1755-1315/183/1/012018

3. Results and Discussion

3.1. Sucrose-concentration in leaf and latex of two rubber clones

The results of sucrose analysis on rubber tree leaves were measured in mM for the sucrose-concentration and in mg g^{-1} for sucrose-content values. The results showed that the sucrose-concentration in the rubber leaves was not influenced by genotype differences in tested clones used in this study. The sucrose-concentrations were measured at 0.052 mM and 0.044 mM in PB 260 and PR 300, respectively (**Figure 2a**). In the rubber tree, the sucrose parameter determined in latex is known to be associated with the biosynthesis of rubber particles. The sucrose-concentration in latex is the difference between sucrose-derived sucrose and the amount of sucrose used for latex metabolism [6, 18, 28]. The value of sucrose-concentration reflects the presence sucrose influx for usage in latex regeneration and/or latex metabolism inactivity where the process of reforming sucrose to rubber particle has not been intensive in laticifers [6, 8]. The study of Lizawati showed that the interaction of the sucrose-concentration in latex, especially at the interaction of the scion and rootstock of PR 300 clone was lower (1,060 mM) than the PB 260 clone (1,820 mM) [32, 33] (**Figure 2b**). Both studies used the anthrone colourimetric method of Dische [31]. This method is commonly used to measure the concentration of sucrose in the latex of rubber trees.

The genotype differences in latex metabolism between PB 260 and the PR 300 clone were first believed to be due to the difference of latex sucrose-concentration. The PB 260 clone is a quick starter clone having a high metabolism of latex production while the PR 300 clone is a slow starter clone with low latex metabolism [34, 35]. Herlinawati and Kuswanhadi [36] also reported that the PB 260 clone has a quick starter clone having high latex production in the first tapping. At the beginning of tapping, this PB 260 clone directly converted the sucrose with or without chemical stimulant for latex biosynthesis [6, 36]. The stimulant treatment tends to reduce the availability of sucrose in the leaves or latex. Logically, it leads to an understanding that the sucrose has been converted into rubber particles [6, 10]. The sucrose-concentration in latex PB 260 clone has decreased by increasing the frequency of tapping [37-39]. In PR 300 clone, both in the leaves and latex, the sucrose-concentration was lower than in PB 260. One hypothesis mentioned that in the slow starter clones, the sucrose is probably used for the formation of bark and stem tissue whose bark is thicker than quick starter ones [12, 35, 40, 41].

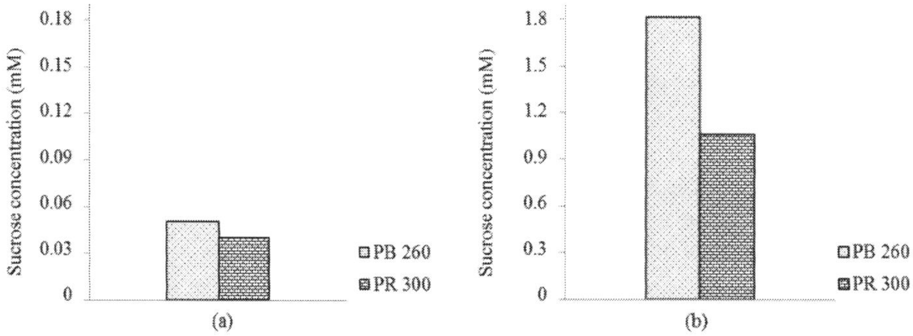

Figure 2. The sucrose-concentration in PB 260 and PR 300 clones (a) leaf (b) latex* (*data from Lizawati [32] and Woelan *et al* [33]).

According to Sumarmadji [6], the high value of sucrose-concentration in rubber tree has not significantly described latex production. It is due to the sucrose-concentration not having a direct correlation from its potential latex productivity. Nevertheless, there is a critical limit on the availability of sucrose-concentrations in laticifers during latex synthesis. Sumarmadji and Tistama [42] stated that the threshold value of sucrose-concentration was about 4 mM. Increasing tapping intensity can reduce the sucrose-concentration value under the threshold so that it will induce the void of latex compiler precursor. Other studies showed similar results for the value of the critical limit of sucrose-concentrations in latex [37]. The sucrose-concentration below 5 mM is a crucial point in latex biosynthesis for determining the pattern of rubber tree planting. The dynamics of sucrose-concentration values in latex were strongly influenced by the metabolic rate inside the whole plant tissues [9]. The range of sucrose-concentration values in mature trees at the beginning of tapping was 3.75-6.91 mM [37]. The low sucrose-concentration indicates the asynchronous metabolism of photosynthesis results in the rubber trees development stage [18, 43]. By contrast, the high sucrose-concentration means less active metabolism. The high concentrations of sucrose can also indicate that the laticifer cells are no longer functioning or degenerating [35, 38, 44]. In this paper, both in leaf and latex, the sucrose-concentration cannot directly describe the actual high production in two contrasted rubber tree clones.

3.2. Sucrose-concentration of two contrasted rubber tree clones under biotic and abiotic stresses

The results of this study showed that the value of sucrose-concentration in the leaf samples varied toward interactions of rubber clone and plant conditions. The value of sucrose-concentrations in WRD-affected trees of PB 260 clones was significantly higher (0.072 mM) compared with healthy (0.046 mM) and TPD-affected trees (0.043 mM) (**Figure 3**). This could be explained by the hypothesis that plant defence system induces a stress signals so that the leaves are forced to perform photosynthesis faster before senescing. WRD causes early yellowing of leaf colour in rubber tree. This mechanism is due to the extraction of the chlorophyll before leaf fall [45]. The low value of sucrose-concentration in TPD-affected trees showed the correlation between sucrose influx and sucrose content in laticifers. Moreover, sucrose is supposed to be used for actively repairing bark tissue previously degraded by TPD. Jacob and Krishnakumar [46] stated that the incidence of TPD occurs due to physiology disturbances in latex biosynthesis and the process after the formation of rubber particles. These mechanisms lead to the anatomy damage of latex vessel tissues.

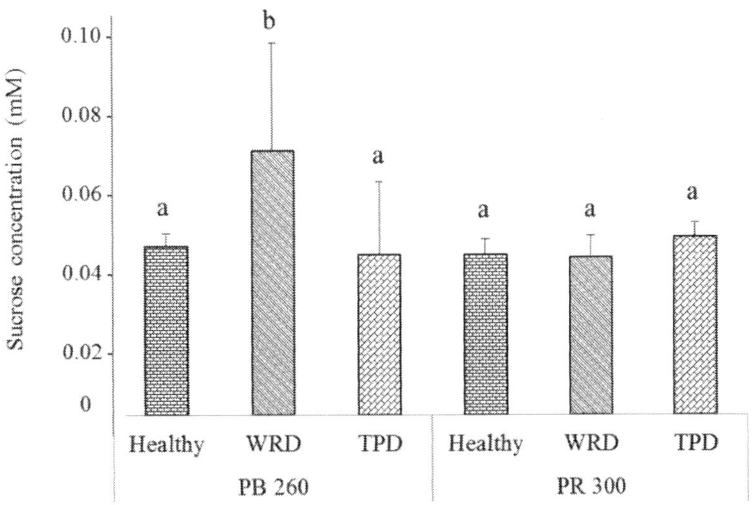

Figure 3. The sucrose-concentration in leaf samples of two contrasted rubber tree clone (PR 260 and PR 300) on healthy-, WRD- and TPD-affected trees condition. WRD: white root disease; TPD: tapping panel dryness. The bars with different letter are significantly different according to the Newman-Keuls test of XL-STAT with the p-value of 0.05.

TPD-affected trees have contained lower macro and micronutrients both in latex and bark compared to healthy trees [47]. In PR 300, the sucrose-concentration values were not significantly different in healthy trees (0.044 mM), WRD-affected trees (0.043 mM) and TPD-affected trees (0.048 mM) (**Figure 3**).

PR 300 having a low latex metabolism in laticifers is more tolerant to TPD. The sucrose is actively used in the recovery process of bark tissue damage. However, the sucrose is not a dominant factor in determining TPD occurrence, since sucrose may not be converted into latex during TPD [6]. According to Gohet *et al.* [48], latex production and developmental aspects such as tree growth actively were actively in competition in the use of sucrose as a precursor of metabolic processes. The allocation of sucrose to laticifer is based on the balance between produced and appropriated sucrose [17, 22].

3.3. Sucrose-content in the leaves of PB 260 and PR 300 clone under WRD and TPD

The content of sucrose in leaves is directly supposed to be equal to the sucrose-concentration value. In PB 260, the sucrose-content in WRD-affected trees was significantly higher (24.45 mg g^{-1}) than healthy (15.53 mg g^{-1}) and TPD-affected trees (13.42 mg g^{-1}) (**Table 2**). These results suggest that the sucrose which has been produced in photosynthesis process is suspected to be in-transmitted into laticifers for latex biosynthesis in WRD-affected trees. In addition, the early infection of WRD could be identified in the leaves damaged by the senescence. Consequently, the leaf-vessel cells do not optimally work. These conditions commonly show changes in the foliage colour, flowering, and prematurely pollination [45, 49, 50]. Tree branches affected by WRD infections will degrade until the entire canopy is destroyed and eventually the rubber tree dies [49, 50]. On the contrary, it can be observed that the sucrose-content in the leaf was relatively low in TPD-affected trees because this physiological disease occurs uniquely in the rubber tree soft bark. The sucrose can still be transported from leaves into laticifers. The PB 260 clone is a high metabolic clone with low sucrose-content in laticifers due to the continuous use of sucrose as a precursor of latex biosynthesis [51, 52]. Thus, PB 260 clone is well known to be susceptible to TPD [47, 53]. In TPD-affected trees, the sucrose is suspected to accumulate higher in the laticifers because it cannot be used as a raw material for latex biosynthesis as the result of latex vessel damage caused by TPD [41, 46, 54, 55].

The high intensity of TPD occurrence is frequently found in clones with high metabolic of latex production such as PB 260 as opposed to low metabolic clones such as PB 217 and PR 300 [41, 53]. The sucrose-content values in PR 300 clone were not significantly different in healthy (14.51 mg g^{-1}), TPD- (16.33 mg g^{-1}), and WRD-affected trees (14.52 mg g^{-1}) (**Table 2**). In a low metabolic clone such as PR 300, the flow of sucrose translocation from the leaf to laticifers is supposedly slow.

Table 2. The sucrose-content value obtained from the calculation of sucrose-concentration standardized by the sucrose standard curve. The value is compared with 0.5 g leaf samples in each treatment.

Rubber tree clones	Rubber tree condition	Sucrose-content (mg g^{-1})	
PB 260	Healthy trees	15.53^{a}	± 1.33
	WRD-affected trees	24.45^{b}	± 9.90
	TPD-affected trees	13.42^{a}	± 6.83
PR 300	Healthy trees	14.51^{a}	± 1.45
	WRD-affected trees	14.52^{a}	± 1.98
	TPD-affected trees	16.33^{a}	± 1.39

*The values with different letter are significantly different according to the Newman-Keuls test of XL-STAT with the p-value of 0.05

* WRD: white root disease; TPD: tapping panel dryness

It also suspected that the sucrose inside laticifers is saturated due to the relatively slow conversion of sucrose to rubber particle in latex biosynthesis [20, 28]. PR 300 clones are moderately tolerant to TPD. Thus, the photosynthesis process to produce sucrose in leaves normally works [38, 39, 56]. The high sucrose-content in the latex indicates good influx in laticifers. But, it can also be the consequence of low metabolism of sucrose conversion into rubber particle [6].

3.4. The clone PB 260 and PR 300 latex yield production

Based on the latex metabolic type, the rubber clones divided into low, medium, medium, high-and medium-high metabolism classes [52]. Commonly, high metabolic clones have higher velocities of latex biosynthesis compared to medium or low metabolic clones. The clonal metabolism type determines the pattern of latex production during tapping exploitation cycles [9, 11, 36, 38]. Physiologically, latex production is influenced by several important factors such as the latex recovery between two tappings and the duration of latex flow [5]. The PB 260 clone has a higher latex production than the PR 300 clone (**Figure 4**). It is caused by the difference in metabolic rate in both of clones to convert sucrose into rubber particles through several enzymatic stages in laticifers. PB 260 is categorized as high metabolic, whereas PR 300 is low metabolic clone to produce latex [35, 39, 57, 58]. The high metabolic rubber tree clones have a high initial production which decreases in mid-age of latex production. In contrast, the low metabolic clone is relatively stable during the whole period of tapping for more than 10 years [35, 39]. The latex productivity of rubber trees depends on several major factors such as the number of laticifers and metabolic activity. The physiological development of laticifers is influenced by the rubber tree clonal genotype and its hormonal interactions [9, 44].

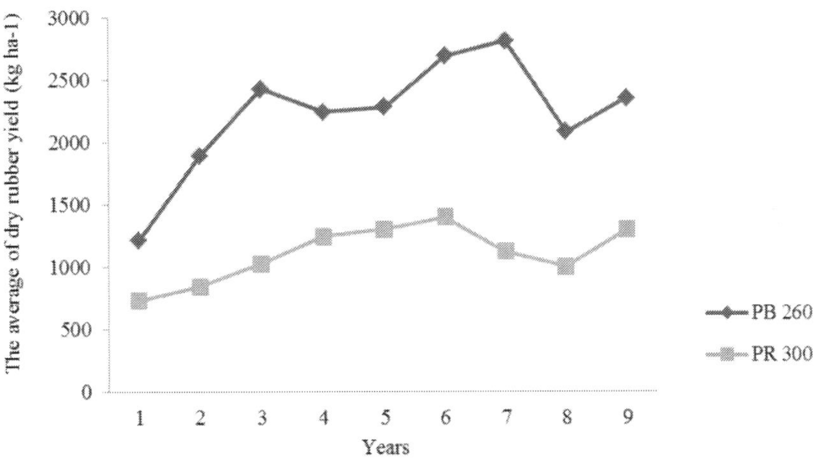

Figure 4. The average of dry rubber yield of quick starter- (PB 260) and slow starter rubber tree clone (PR 300 and PR 300) during 9 years of tapping production [59, 60].

4. Conclusions

The sucrose-content is equivalent to the concentration value of sucrose in the leaves of the rubber tree. The genotype differences between two contrasted clones related to the latex metabolism did not affect the content and concentration value of sucrose in leaves. The interaction between the two rubber clones (PB 260 and PR 300) and the stress conditions (WRD- and TPD-affected trees) showed that only WRD-affected trees influenced the sucrose-content in the leaves of PB 260. This condition was not met in PR 300.

5. Acknowledgements

The authors thank Marini for its technical assistance in the laboratory.

6. References

[1] Castelblanque L, Balaguer B, Martí C, Rodríguez J J, Orozco M and Vera P 2016 *Plant Physiol.* **172** 1032
[2] de Faÿ E and Jacob J 1989 Anatomical organization of the laticiferous system in the bark *Physiology of Rubber Tree Latex* ed d'Auzac, Jacob J, Chrestin H Boca (Florida (US): CRC Press Inc) p. 4-14
[3] d'Auzac J and Jacob J 1989 The composition of latex from *Hevea brasiliensis* as a laticiferous cytoplasm *Physiology of Rubber Tree Latex* ed d'Auzac, Jacob J, Chrestin H Boca (Florida (US): CRC Press Inc) p. 58-96
[4] Kongsawadworakul P and Chrestin H 2003 *Plant and Cell Physiol.* **44** 707
[5] Tian W M, Zhang H, Yang S G, Shi M J, Wang X C and Dai L J 2013 *J. Plant Physiol.* **170** 723
[6] Sumarmadji 1999 The response of physiological characters and latex production of some rubber tree clones to ethylene stimulation [*Dissertation*]. Bogor (ID): Bogor Agricultural University
[7] Priyadarshan P 2017 *Biology of Hevea Rubber* (Kerala (IN): Springer International Publishing)
[8] Atminingsih A, Napitupulu J A and Siregar T H 2016 *Indonesian J. Nat. Rubber Res.* **34** 13
[9] Jacob J L, Prévôt J C, Roussel D, Lacrotte R, Serres E and d'Auzac J 1989 Yield limiting factors, latex physiological parameters, latex diagnosis, and clonal typology *Physiology of Rubber Tree Latex* ed d'Auzac, Jacob J, Chrestin H Boca (Florida (US): CRC Press Inc) p 345-382
[10] Sumarmadji 2004 *J. Rubber Res.* **22** 70
[11] Jacob JL, Eschbach JM, Prévot J, Roussel D, Lacrotte R, Chrestin H 1986 *International Rubber Conference* Kuala Lumpur (MY): Rubber Research Institute of Malaysia
[12] Silpi U, Thaler P, Kasemsap P, Lacointe A, Chantuma A and Adam B 2006 *Tree Physiol.* **26** 1579
[13] Tang C, Huang D, Yang J, Liu S, Sakr Sand Li H 2010 *Plant, Cell Environ.* **33** 1708
[14] Tian W M, Yang S G, Shi M J, Zhang S X and Wu J L 2015 *J. Plant Physiol.* **182** 95
[15] Silpi U, Lacointe A, Kasempsap P, Thanysawanyangkura S, Chantuma P and Gohet E 2007 *Tree Physiol.* **27** 881
[16] Eschbach J M, Tupý J and Lacrotte R 1986 *Biol. Plant.* **28** 321
[17] Tupy J 1973 *Physiologie végétale*
[18] Tupý J 1989 Sucrose supply and utilization for latex production *Physiology of Rubber Tree Latex* ed d'Auzac, Jacob J, Chrestin H Boca (Florida (US): CRC Press Inc) p. 179-218
[19] Tupý J, D'Azuac J and Chrestin H 1988 *Physiology of rubber tree latex*
[20] Jacob J L, Prevôt J C and D'Auzac J 1982 *Phytochem.* **21** 851
[21] Jacob J, Serres E, Prévôt J, Lacrotte R, Vidal A and Eschbach J 1988 *Agritrop.* **12** 95
[22] Jacob J L and Prévôt J C 1992 *Developments in Crop Sci.* **23** 116
[23] Chow K S, Mat-Isa M N, Bahari A, Ghazali A K, Alias H and Mohd.-Zainuddin Z 2011 *J. Experiment. Bot.* **63** 1863
[24] Chow K S, Wan K L, Mat-Isa M N, Bahari A, Tan S H and Harikrishna K 2007 *J. Experiment. Bot.* **58** 2429
[25] Mesquita A C, Oliveira L E, Mazzafera P and Delú-Filho N 2006 *Brazilian J. Plant Physiol.* **18** 263
[26] Dusotoit-Coucaud A, Brunel N, Franchel J, Granet F, Chrestin H and Sakr S 2009 *J. Rubber Res.* **12** 125
[27] Dusotoit-Coucaud A, Brunel N, Kongsawadworakul P, Viboonjun U, Lacointe A and Julien J L 2009 *Ann. Bot.* **104** 6357

[28] Silpi U, Chantuma P, Kasemsap P, Thaler P, Thanisawanyangkura S and Lacointe A 2006 *J. Rubber Res.* **9** 115
[29] Zhu J, Qi J, Fang Y, Xiao X, Li J and Lan J 2018 *Frontiers in Plant Sci.* **9** 58
[30] Whittaker A, Martinelli T, Farrant J M, Bochicchio A and Vazzana C 2007 *J. Experiment. Bot.* **58** 3775
[31] Dische Z M 1962 *Methods in Carbohydrate Chemisrty* ed Whistler R L, Wolfrom, M L (New York (US): Academic Press
[32] Lizawati 2009 *Indonesian J. Agron.* **13** 19
[33] Woelan S, Sayurandi S and Pasaribu S A 2013 *J. Rubber Res.* **31** 1
[34] Siregar T and Suhendry I 2013 *Cultivation and Rubber Technology* (Jakarta (ID): Penebar Swadaya p.236
[35] Sumarmadji, Karyudi, Siregar T 2006 *Proc Loc Nat. Plantation of Rubber Trees* (Medan (ID): Indonesian Rubber Research Institute) p.4-6
[36] Herlinawati E and Kuswanhadi 2013 *J. Rubber Res.* **31** 110
[37] Koryati T 2016 The efforts to speed up mature tapping and latex production characters of rubber clones through the use of growth regulating substances [*Dissertation*] (Medan (ID): USU)
[38] Kuswanhadi, Sumarmadji, Karyudi and Siregar T 2009 *Proc Loc Nat Plantation of Rubber Trees* (Batam (ID): Indonesian Rubber Research Institute) p.150-156
[39] Sumarmadji S, Rouf A, Aji Y B S and Widyasari T 2017 *Warta Perkaretan* **36** 55
[40] Silpi T and Lacointe A 2006 *J. Rubber Res.* **9** 2
[41] Sumarmadji 2005 *J. Rubber Res.* **23** 58
[42] Sumarmadji and Tistama R 2004 *J. Rubber Res.* **22** 27
[43] de Souza G A, dos Santos Dias D C F, Pimenta T M, Almeida A L, de Toledo Picoli E A and de Pádua Alvarenga A 2018 *Physiol. Plant.* **162** 495
[44] Tistama R, Lubis V and Nurwahyuni I 2017 *Buletin Anatomi dan Fisiologi* **2** 1
[45] Wattanasilakorn S, Sdoodee S, Nualsri C, Chuenchit S, Meesawat U and Sopharat J 2016 *Walailak J. Sci. Technol. (WJST)* **14** 549
[46] Jacob J and Krishnakumar R 2006 Tapping panel syndrome: what we know and what we do not know *Tapping Panel Dryness Of Rubber Trees* ed Jacob J, Krishnakumar R, Mathew NM *International Workshop on Tapping Panel Dryness* Kerala (IN): Rubber Research Institute of India p.17-27
[47] Andriyanto M and Tistama R 2014 *Warta Perkaretan* **33** 89
[48] Gohet E, Clement A, Eschbach J, Jacob J and Prevot J 1996 *Plantations, Recherche, Développement* **3** 30
[49] Omo-Ikerodah E, Omorusi V and Mokwunye M 2012 *World Rural Obs* **4**
[50] Omorusi V I 2012 *Plant Science: InTech*
[51] Clément-Demange A, Priyadarshan P, Thuy Hoa T T and Venkatachalam P 2007 *Plant Breeding Rev.* **29** 177
[52] Gohet E, Chantuma P, Lacote R, Obouayeba S, Dian K and Clément-Demange A 2003 Latex clonal typology of *Hevea brasiliensis*: Physiological modelling of yield potential and clonal response to ethephon stimulation. *Workshop on Exploitation Technology* Kottayam (IN): IRRDB
[53] Tistama R, Sumarmadji and Siswanto 2006 *Proc Loc Nat Plantation of Rubber Trees* (Medan (ID): Indonesian Rubber Research Institute)
[54] Abraham T, Mathew J, Srinivas P and Jacob C K 2006 Incidence of tapping panel dryness on popular rubber clones in southern rubber growing region of India *Tapping Panel Dryness Of Rubber Trees* ed Jacob J, Krishnakumar R, Mathew NM *International Workshop on Tapping Panel Dryness* Kerala (IN): Rubber Research Institute of India p.1-16
[55] Gomez J B, Samsidar H, Ghandimathi H and Ho L H 1990 *J. Nat. Rubber Res.* **5** 90
[56] Boerhendhy I and Amypalupy K 2016 *J. Agric. Res. Dev.* **30** 23

IOP Conf. Series: Earth and Environmental Science **183** (2018) 012018 doi:10.1088/1755-1315/183/1/012018

[57] Siagian N 2012 *Pembibitan dan Pengadaan Bahan Tanam Karet Unggul* ed Karyudi (Medan(ID): Balit Sungei Putih p.117

[58] Siregar T H S, Junaidi, Sumarmadji, Siagian, N and Karyudi 2008 *Proc Loc Nat Rubber Agribusiness* Yogyakarta(ID): Indonesian Rubber Research Institute p. 217-232

[59] Madjid A, Angreani-Hendranata, Saleh M and Soeselo S 1975 Result of clone valuation *Menara Perkebunan* **43** 235

[60] Woelan S, Aidi-Daslin, Lasminingsih M and Suhendry I 2009 Evaluation of IRR rubber clone series 200 and 300 during the trial stage *Proc Loc Nat Rubber Breeding* Batam (ID): Indonesian Rubber Research Institute p. 236-244.

IOP Conf. Series: Earth and Environmental Science **183** (2018) 012019 doi:10.1088/1755-1315/183/1/012019

Chemical and biological characteristics of selected wet soil in oil palm plantation attack with *Ganoderma* sp.

Happy Widiastuti[1*], D Taniwiryono[1], H Hendarjanti[2], S Harjotedjo[2] and M Hadi Sugeng W[2]

[1]Indonesian Research Institute for Biotechnology and Bioindustry, Jalan Taman Kencana No.1, Bogor 16128, Indonesia
[2]PT Astra Agro Lestari Tbk, Jalan Pulo Ayang Raya Blok OR 1, Kawasan Industri Pulo Gadung Jakarta Timur

*E-mail: happywidiastuti@yahoo.com

Abstract. The expansion of oil palm plantation requires the use of suboptimal land such as peat soil. *Ganoderma* as soil born pathogen is the most important pathogen in palm oil cultivation. The objective of the study was to reveal the characteristics of chemical and biological characteristics of wet soil on land with mild and severe *Ganoderma* attack. The study was conducted in three oil palm estates with wet soil at two levels attack of *Ganoderma* sp. Palm oil is the first-generation plant and the previous habitat was a forest. The soil samples were taken from three principal plants and each staple, taken from four sample points, namely harvesting path, inter-row, around the stem, and the ground near the base of the stem. The soil samples were analyzed for the nutrient status of N, P, K, Mg, Cu and C by standard methods and soil biology characteristics such as total bacteria, total fungi, cellulolytic fungi, and *Azotobacter*. The results showed that there was a correlation between soil nutrient status of K and Mg with *Ganoderma* sp. attack on all of the plantation tested. While for Cu and C, there was a correlation between soil nutrient status and *Ganoderma* attack in 3 plantations only. The least correlation is between N and *Ganoderma* sp. While the correlation between total bacteria with *Ganoderma* sp. was most common although very low.

1. Introduction

Various soil management practice such as in balance of chemical fertilizer doses application and no addition of organic fertilizer supposed increase the incidence of basal stem rot disease caused by *Ganoderma* sp. The use of chemical fertilizers for many years and there is no practice of returning biomass to the soil, in turn, leads to disruption of the biotic soil communities. This imbalance, impacts on the decreasing of some soil properties and one of this suppressed soil microbial communities controlling soil borne pathogens such as *Ganoderma* sp., a fungus that causes major diseases in oil palm.

The soil can be said to be a major terrestrial ecosystem organizer. Soil minerals, organic components and soil microorganisms cannot be classified as separate parts of the constituent components of the soil but as a single unit that continuously maintains interaction and unity of one another in the terrestrial environment. The interaction of soil system components can affect the number of soil microbes, the global ionic cycle of soils, global climate change, biodiversity, biological

IOP Conf. Series: Earth and Environmental Science **183** (2018) 012019 doi:10.1088/1755-1315/183/1/012019

production and nutrients for humans derived from nature. In general, land/land degradation is defined as a decrease in the value of soil/function of the soil as a result of the decreasing of important elements of soil constituents or also as a decrease in potential/usefulness and decreased the ability to support the soil ecosystem [1]. This leads to a decrease in biomass production as well as the ability of water and air plant management [2]. Soil degradation is divided into 3 types of degradation processes namely physical degradation, chemical degradation, and biological degradation. Degradation of the soil will physically affect the structure of the soil, the ability of the soil to absorb water and air, and resistance to destruction by the flow of water and air. Soil chemical degradation affects soil acidity (pH), decreases availability and ease of use of nutrients for plants, the ability to destroy toxins for other organisms, and decreases the excessive increase of salt levels in plant root zones. Biological degradation affects the availability of soil organic carbon (SOC), the diversity of soil biota species and increases the population of soil pathogens [3].

The results showed that the main problem of wetland is bad aeration and anaerobe condition that does not support the development of plant roots. Restructuring of wetland can be done with application of empty fruit bunches oil palm which can be combined with soil trenching making so it is expected to improve the characteristics of the land include improvement of water conservation, plant roots so as to increase nutrient uptake, and wet soil aeration.

This paper aims to reveal the correlation between soil characteristics of both biological and chemical and the *Ganoderma* sp. attack. The results of this study, expected to know the relationship of soil character with *Ganoderma* as a fungus with soil as a growing habitat. In the future, it is hoped that improvement of soil character could be used as one technique in controlling BSR disease caused by *Ganoderma* sp.

2. Materials and Methods

The research was conducted by taking samples from four gardens i.e. Mamuang, Pasang Kayu, Letawa and LTT (**Table 1**).

Table 1. Location of samples used in this research.

Estate	Block	Year plant	Area (ha)	Yield	*Ganoderma* level attack (%)	Classification
Mamuang	A13	1997	29.99	29.17	0.0060	Low
	K7	1996	38.95	25.72	0.0070	Low
	G14	1995	07.77	35.00	0.0400	Severe
	G15	1997	39.09	24.40	0.0300	Severe
Pasang	C21	2002	20.93	17.04	0.0000	Low
Kayu	E16	2002	37.62	22.04	0.0700	Low
	E8	1996	39.73	26.86	1.2200	Severe
	E5	1998	41.17	24.39	1.4500	Severe
Letawa	F15	1999	44.00	23.84	0.0022	Low
	L5	2000	34.88	26.26	0.0094	Low
	L3	1998	31.46	23.62	0.0250	Severe
	L9	1995	31.22	21.23	0.0330	Severe

2.1. Analysis of soil and palm oil plant tissue

Leaf and soil nutrient and microbial diversity analysis were conducted to find out the general description of plant nutrient status with *Ganoderma* attack. The soil sampling for nutrient and microbial analysis is taken from 3 points in 1 block consisting of the 1st point (main point), the 2nd point is the 6th lane from the main point (towards the West), and the point 3 is the 12th line from the main point (towards the South). The sample is taken from a healthy tree and not located on the edge of the area. At each point, 4 soil samples were taken from harvesting path, inter-row, around the stem, and the ground near the base of the stem. Then all four samples were composite to be one sample. Soil sampling is taken by using bamboo. The depth of soil taken is 20 cm from the soil surface and the weight of samples taken is 500 g. Microbial analyses included total fungi, cellulolytic fungi, bacteria, N-fixing bacteria (*Azotobacter* sp.), while the nutrients analyzed included C, N, P, K, Mg, and Cu. Leaf samples were taken from the same location as the soil in the form of 5 pieces of the leaves the right and left after the sweet spines. The leaf nutrients analyzed was N, P, K, and Cu.

3. Results and Discussion

3.1. Analysis of soil and palm oil plant tissue

3.1.1. Chemical characteristics of soil

The correlation of chemical properties of soil in Mamuang gardens with *Ganoderma* sp. attack are presented in **Table 2**. The results of chemical analysis or chemical characteristics are not distinguished between sampling sites i.e. base, circle, and "harvesting path". The carbon content in Mamuang soil is highly correlated with *Ganoderma* sp. The higher the carbon content the lower the attack rate of *Ganoderma* sp. This result can be understood because by increasing the carbon content of the population of heterotrophs and especially the antagonist's microbe to *Ganoderma* sp. and suppress the activity and development of *Ganoderma* sp. However, it is suspected that the form of carbon compound is more correlated with the attack rate of *Ganoderma* sp. compared with the total carbon compounds. Qudsai *et al* [4] showed that there is positively and significantly correlated between pathogen with soil organic matter and nitrogen content. Kanwal *et al* [5] showed that in the rice-wheat cropping system, number of saprophytic fungi colonies and the total number of fungal colonies were significantly and negatively correlated with organic matter and N, but positively correlated with EC and K, respectively. Correlation between a number of colonies of pathogenic fungi and other soil characteristics was insignificant.

A high correlation was also found between K_2O content and *Ganoderma* sp. To some extent, K content will decrease *Ganoderma* attack but in certain content will increase *Ganoderma* sp. attack. The low content of K will increase pathogenic infection both obligate and facultative. Potassium is a macronutrient that plants need for membrane permeability. A high correlation was also found between MgO content and *Ganoderma* sp. The equation describing the relationship between the two is the quadratic which means the higher the MgO the higher the attack *Ganoderma* sp. Marschner [6] suggests that Mg is a macronutrient needed by plants as a chlorophyll component.

A high correlation was also found between Cu content and *Ganoderma* sp. from the equation, the relationship between the Cu content and the attack rate of *Ganoderma* sp. shown that the higher soil Cu content will be followed by increasing attacks of *Ganoderma* sp. Cu is a cofactor for laccase enzyme [7] secrete by *Ganoderma* sp. that belong to the white-rot Basidiomycetes. This possibility is related to laccase activity of *Ganoderma* sp. Nevertheless, the linkage of the Cu content of soil with an attack rate of *Ganoderma* sp. needs to be tested further.

The correlation analysis of some soil chemical properties in the Pasang Kayu shows that there is a high correlation between Carbon, K_2O, MgO and Cu with the attack rate of *Ganoderma* sp. (**Table 3**). The high correlation of K_2O, MgO and Cu with the attack rate of *Ganoderma* sp. seems to be the same as those found in Mamuang and Letawa.

Table 2. Correlation of soil chemical characteristics and *Ganoderma* sp. attack of Mamuang estate.

Variable parameter	Functional equation	Coefficient determination R^2
N total	$Y=-1.543x^2+2.5x+1.364$	0.474
C total	$Y=-0.036x+2.674$	0.972
K_2O total	$Y=-556x^2+88.14x-1.329$	0.911
P_2O_5 total	$Y=-6.59x^2-0.034x+1.854$	0.156
MgO total	$Y=-0.538x^2+1.951x-0.260$	0.982
Cu	$Y=0.059x+0.116$	0.820

Table 3. Correlation of soil chemical characteristics and *Ganoderma* sp. attack of Pasang Kayu.

Soil character	Equation	R^2
N total	$Y=1.519x^2-2.134x+1.925$	0.129
C total	$Y=-0.001x^2+0.098x+0.324$	0.999
K_2O total	$Y=267.1x^2-44.53x+2.813$	0.870
P_2O_5 total	$Y=11.18x^2-12.61x+3.225$	0.505
MgO total	$Y=0.847x^2-2.144x+2.274$	0.975
Cu	$Y=0.001x^2-0.117x+3.295$	0.944

Analysis of soil carbon content, N, K, Mg and Cu of Letawa showed a fairly high correlation with *Ganoderma* sp. (**Table 4**). The equation which illustrates the fourth correlation of the soil with *Ganoderma* sp attack is quadratic. These results indicate that at certain nutrient levels the four nutrients will reduce the attack *Ganoderma* sp.

Table 4. Correlation of soil chemical characteristics and *Ganoderma* sp. attack of Letawa.

Soil characteristics	Equation	R^2
N total	$Y=-5.135x^2+6.297x+0.418$	0.827
C total	$Y=-0.011x^2+0.355x+0.619$	0.846
K_2O total	$Y=-152.4x^2+39.99x-0.24$	0.879
P_2O_5 total	$Y=136.1x^2-7.207x+1.107$	0.460
MgO total	$Y=-2.3x^2-3.301x+0.846$	0.984
Cu	$Y=-0.0x^2+0.055x+7.34$	0.985

The results of the chemical analysis showed that the correlation between MgO, K_2O, N and P_2O_5 in LTT soil correlated with the attack rate of *Ganoderma* sp. (**Table 5**). The equation which illustrates the fourth correlation of the soil with *Ganoderma* sp. attack is quadratic. The high correlation of K_2O, N, P, and MgO with the attack rate of *Ganoderma* sp. seems to be the same as those found in PSK, Mamuang and Letawa.

IOP Conf. Series: Earth and Environmental Science **183** (2018) 012019 doi:10.1088/1755-1315/183/1/012019

Table 5. Correlation of soil chemical characteristics and *Ganoderma* sp. attack of LTT.

Soil character	Equation	R^2
N	$Y=-14.34x^2+10.9x+0.208$	0.887
C	$Y=-0.003x^2-0.127x+2.225$	0.434
K_2O	$Y=-36.41x^2+15.4x+0.404$	0.909
P_2O_5	$Y=69.87x^2-32.03x+4.514$	0.790
MgO total	$Y=0.41x^2+0.387x-0.88$	0.985
Cu	$Y=-0.033x^2+1.682x-19.36$	0.105

3.1.2. The characteristic of leaf nutrient content of palm

The analyses of leaf nutrient are presented in **Table 6**. It is shown that leaf nutrient content in all blocks did not show any association between intensity of *Ganoderma* sp. attack and leaf nutrient content. However, this is thought to be due to a very limited number of examples.

Leaf N content ranges between 2.43% (Pasang Kayu, heavy) - 3.13% (LTT, weight) and based on literature studies, the levels of N in this range are sufficient. Leaf P analysis showed a range between 0.19 (Mamuang, height) - 0.24% (Mamuang, low). The range of normal P content is between 0.3 - 0.5%. The result of P analysis shows that the leaf P content is below the normal or low range. Marschner (1986) suggested that P deficiency would cause hormonal imbalances, plant growth inhibition, chlorophyll content increase despite low photosynthetic efficiency. The leaf K content of the analysis also showed deficiency because the normal K content was in the range of 2 - 5%. In the analyzed samples the lowest content (0.78%) i.e. the leaves in Pasang Kayu with severe *Ganoderma* attack, while the highest content (1.91%) was leaves in LTT with severe *Ganoderma* attack too. The deficiency of K elements causes the plant to be easily infected with both obligate and facultative parasite fungi. This is because, at low K, the synthesis of high molecular weight organic compounds decreases and accumulates low molecular weight organic compounds.

Tabel 6. Soil chemical characteristics of oil palm leaf

Estate	Leaf nutrient content				
	***Ganoderma* attack (%)**	**N (%)**	**P (%)**	**K (%)**	**Cu (ppm)**
Mamuang	Severe	2.66	0.21	0.96	3.60
Mamuang	Low	2.68	0.24	1.49	4.10
Mamuang	Low	2.66	0.22	0.96	2.60
Mamuang	Severe	2.50	0.19	1.02	4.50
Pasang Kayu	Low	2.58	0.21	1.43	4.10
Pasang Kayu	Severe	2.43	0.20	0.78	4.90
LTT	Low	2.72	0.21	1.25	8.80
LTT	Severe	3.13	0.24	1.91	8.70
Letawa	Low	2.63	0.20	1.30	8.30
Letawa	Severe	2.87	0.22	1.45	9.40

The leaf content of Cu leaves is in the normal range, whereas for leaf samples from Mamuang, one deficient example (less than 3%) detected low *Ganoderma* attack. Nevertheless, leaf samples from LTT and Letawa have a higher Cu content above normal. Cu nutrient which is a micronutrient

IOP Conf. Series: Earth and Environmental Science **183** (2018) 012019 doi:10.1088/1755-1315/183/1/012019

function in the process of lignification of the cell wall of plants so that plants that have Cu deficiency cell wall damage, containing cellulose high while low lignin. However, this element is also a cofactor of the laccase enzyme, which is one of the enzymes possessed by *Ganoderma* that plays a role in the delignification of cell walls. However, the plant will interact with *Ganoderma* sp. and nutrient status will affect the anatomy and physiology of the host plant and the activity of *Ganoderma* sp. Management of plant nutrient status that supports plant growth but suppresses pathogenic activity is very important.

3.1.3. Characteristics of soil biology in experimental land

The result of correlation analysis between several microbial function and attack rate of *Ganoderma* sp. in Mamuang showed that the total fungi, total bacteria and population of *Azotobacter* had low correlation with *Ganoderma* sp. However, for the population of cellulolytic fungi showed a rather high correlation number that is 0.460. While Nuria *et al* [8] showed that manipulating the soil microorganism community leading to suppress the disease.

Similar results were also found in the soil of Letawa which is only total bacteria that have the highest correlation with *Ganoderma* attack that is 0.493. This result is the same as the soil sample in LTT but the correction is 0.370. In contrast to Mamuang, Letawa and LTT, the correlation between total bacteria and soil fungi Pasang Kayu with *Ganoderma* sp. attacks i.e. 0.700 and 0.540 respectively. De Graaff *et al* [9] reported that there is a correlation between soil biodiversity and the soil C cycling process. It also highlights the importance of diversity across groups of organisms (e.g., primary consumers and secondary decomposers) for maintaining full functionality of C cycle processes. Moreover, the statistical analysis of the DGGE fingerprints showed that both plant age and genotypes influenced the bacterial community structure in the tuber rhizosphere [10].

3.1.4. The principal census of Ganoderma *attacks on the ground.*

Based on the criteria of the principal census of *Ganoderma* attacked the land, it was found that in Mamuang there were 8 palms with yellow status, 5 red, and 105 items with a green status. Data on Pasang Kayu, there was 15 palm trees with yellow status, 54 palm trees with red status and 88 green colours, whereas in Letawa there are 11 trees with yellow status, 51 trees with red status and 236 palm trees with a green status. The oil palm tree with white status in all land not found. With these data then the existence of *Ganoderma* sp. should be various. The observations with the highest number of attacks were found in the Pasang Kayu then Letawa and Mamuang. Based on previous field research results it is known that if in one hectare with planting density 130 trees there are 20 dead plants, then usually on the land 10 plants have been heavily infected, and 10 plants have been infected lightly. The productivity of 10 plants with a heavy infestation is up to 50%. The loss of one-hectare land production can be calculated by the formula ([20 + (50% x 10)] / 130) x 100% equals 19.23% [9]. Thus, it is appropriate that the field of research is treated with preventive measures *Ganoderma* sp. by increasing the consortium of superior microbes such as *Trichoderma* sp., N fixing bacteria, bacterial solubilized phosphate, mycorrhiza and combination of them.

4. Conclusions

Research sites are mostly wet soils. Some soil nutrients correlate with *Ganoderma* attacks. The correlation between microbial populations with *Ganoderma* attacks is very low. The results of leaf nutrient analysis showed that P and K nutrient deficiency was present in all instances of oil palm leaf. While in the Letawa and LTT, the analyses showed that there is the excessive content of Cu leaf.

However, the data were only drawn from a limited number of samples taken from a very limited area and only from individual plants with two levels of *Ganoderma* sp. attack. i.e. light and heavy. Further analysis needs to be done to determine the nutrient status of individual plant samples associated with disease progression in individual plants of the same sample over time. In addition, it is necessary to examine the relationship of nutrient status in the garden with mild attack rate (less than 1%), moderate (about 5%), and weight (about 30%) in the wider area.

5. References

[1] Lal R 2004 Science **304** 1623
[2] Bardgett R, C Freeman and N J Ostle 2008 *Int. Soc. Micro. Ecol. J.* **2** 805
[3] Lal R 2010 *BioScience* **60** 708
[4] Qudsai H, A Javaid, R Mahmood and N Akhtar 2017 *Pak. J. Bot.* **49** 1579
[5] Kanwal A, J R Mahmood and N Akhtar 2017 *The J. Animal & Plant Sciences* **27** 1256
[6] Marschner H 1986 *Mineral Nutrition in Higher Plants* (London: Acad. Press)
[7] More SS, P S Renuka, K Pruthvi, M Swetha, S Malini S and S M Veena 2011 *Enzyme Res.* PMCID: PMC3184503 PMID: 21977312
[8] Nuria B, J A Gutiérrez-Barranquero, A de Vicente and F M Cazorla 2012 *Diversity* **4** 475
[9] de Graaff, J Adkins, P Kardol, HL Throop 2015 *SOIL* **1** 257
[10] Marques J M., da Silva T F, Vollu R E, Blank A F, Guo-Chun Ding, Seldin L and Smalla K 2014 *FEMS Microbiol Ecol* **88** 424
[11] Darmono 2010 Personal communication

IOP Conf. Series: Earth and Environmental Science **183** (2018) 012020 doi:10.1088/1755-1315/183/1/012020

Modified *Ganoderma* selective medium to meet Indonesia's government regulation

Wiwit Iman Amanda[1] and H T Prakoso[2]*

[1]Department of Agroecotechnology, Faculty of Agriculture, University of Brawijaya, Jalan Veteran 1, Malang, 65145, Indonesia
[2]Indonesian Research Institute for Biotechnology and Bioindustry, Jalan Taman Kencana No.1, Bogor 16128, Indonesia

*E-mail: haryo.tejo@iribb.org

Abstract. *Ganoderma boninense*, the causal agent of basal stem rot in oil palm, has caused great losses on oil palm industry. Isolation method of the fungus from the field for further study is considered problematic due to high contaminations. To tackle this, *Ganoderma* selective medium (GSM) has been developed by scientists from Malaysia. The problem occurred when the media is going to be used in Indonesia due to the prohibition of one of its ingredients by Indonesia's government regulation No. 85/1995 on the management of hazardous wastes and toxic. This study aims to test whether the selective medium is still effective even without Pentachloronitrobenzene (PCNB) and Ridomil. The former is the prohibited ingredients and the latter is not well distributed in Indonesia thus considered as a rare item. The study showed that contamination can be kept to minimal and the distinguished feature of GSM, formation of a brown halo around the colony, was seen. However, the growth of *Ganoderma*'s mycelium seems hampered. We hypothesize that besides being a fungicide, PCNB may play a role as a carbon source that supports growth for *Ganoderma*. The find for substitute ingredient of PCNB in GSM is on its way.

1. Introduction

Ganoderma boninense, the causal agent of basal stem rot in oil palm, has caused great losses as much as IDR 3.6 billion on oil palm industry in Indonesia [1]. This disease has become 'cancer' in the oil palm industry due to its persistence and sometimes asymptomatic. The fungus can live in the soil for a long period of time, as long as its substrates, such as root debris or decayed root, are still present in soil [2]. Based on field observation on an area which has been undergone crop rotation from oil palm to sugarcane for 5 years, young oil palm seedling that grows from left-over oil palm fruit during harvesting period shows symptoms that indicate *Ganoderma boninense* is still present. This observation supports Hasan and Turner [3] findings that the inoculum was still viable in the ex-oil palm-planted-soil even though fallow period occurred for many years.

Isolation of this fungus from the field is also problematic due to high contaminations during the process. Ariffin, Idris and Singh [4] came up with a solution of formulating a selective media that can distinguish *Ganoderma* among others. However, *Ganoderma* Selective Medium (GSM) may not be used in Indonesia due to the restriction of one from its ingredients, pentachloronitrobenzene, which considered hazardous and toxic as stated by Indonesia's government regulation No. 85/1999 on Toxic and Hazardous Waste Management [5]. According to a review from Arora and Bae [6], PCNB may cause liver damage, renal agenesis, and hepatomas on tested animals and thus PCNB is considered as carcinogenic by the United States Environmental Protection Agency (EPA). Moreover, another GSM's ingredient namely Ridomil has not distributed anymore in Indonesia. Therefore, this study aims to test whether the selective medium was still effective even without Pentachloronitrobenzene (PCNB) and Ridomil and to check alternative ingredient that can substitute PCNB

2. Materials and Methods

2.1 GSM Preparation

GSM consists of two parts: part A and part B. The original composition of GSM is showed in **Table 1**.

Table 1. GSM composition [4].

Ingredients	Amount
Part A	
Bacto-Peptone	5 g
Agar	20 g
K_2HPO_4	0.5 g
Distilled water, pH 5.5	900 ml
Part B	
Streptomycin Sulphate	300 mg
Chloramphenicol	100 mg
PCNB, pure	285 mg
Ridomil (25% WP)	130 mg
Benlate – T20	150 mg
Ethanol, 95%	20 ml
Lactic acid, 50%	2 ml
Tannic acid	1.25 g
Distilled water, pH 5.5	80 ml

Part A was homogenized by stirring it on a hot plate until dissolved with a temperature of 100 °C followed by autoclaved for 15 minutes. Meanwhile, part B was stirred for two hours at room temperature. Part B is aseptically poured on Part A when the temperature of part A has decreased to around 45 °C.

2.2 Selectivity test of GSM without PCNB and Ridomil towards Ganoderma

Isolate *Ganoderma boninense* AR77 and B23 originated from oil palm field in Riau and Lampung, respectively, were used in this study. To check GSM selectivity, other fungi were used, namely *Fusarium* sp., *Omphalina* sp., *Pleurotus* sp., and *Trichoderma* sp. The isolates were rejuvenated by growing it on Potato Dextrose Agar (PDA) and was let sit in the dark for 7 days. The edge of the freshly growing hyphae was then cut and aseptically transferred to GSM without PCNB and Ridomil and put in the dark for another 7 days. At the end of incubation, the formation of the brown halo was checked.

2.3 Selectivity test of GSM added with guaiacol towards Ganoderma.

Guaiacol was used to test whether it can function as a substitute ingredient for PCNB. The method for preparation of GSM+guaiacol as well as fungi inoculation, incubation and the observation of brown halo were not different from GSM without PCNB and Ridomil. The only difference was the addition of 1.2 ml Guaiacol to Part B of GSM.

3. Results and Discussion

3.1 Selectivity test of GSM without PCNB and Ridomil

As can be seen from **Figure 1**, brown halo was formed around *Ganoderma* colony due to its specific reaction with tannic acid, but not on the other ligninolytic fungi [7]. Tannic acid is also functioned as an inhibited growth compound for contaminant due to its antibacterial activities [8]. The hyphae of *Ganoderma* seemed did not grow well on modified GSM (without PCNB).

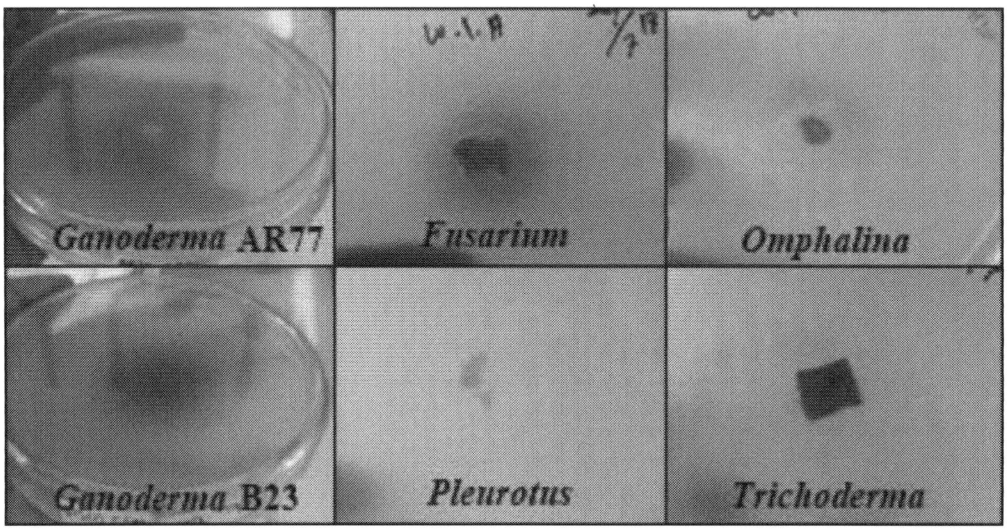

Figure 1. The morphology of the fungi that were grown on GSM without PCNB and Ridomil. Noticed that only Ganoderma that gave the formation of brown halo around the agar.

GSM selectivity is based on the formation of brown halo when *Ganoderma* grow upon the medium [4]. Moreover, it would inhibit other bacteria or fungi growth because there are abundant of anti-bacteria and anti-fungi that are mixed in the media. These characteristics were still can be seen on modified GSM. However, *Ganoderma* seemed cannot grow normally on modified GSM. Its hyphae are not growing even when the isolate was kept in dark chamber for 30 days. When *Ganoderma* from modified GSM was transferred to PDA, the hyphae could grow normally (**Figure 2**). Therefore, it is suggested that PCNB might act as a carbon source in GSM.

Figure 2. *Ganoderma* can grow normal when transferred from modified GSM to PDA.

Although PCNB is widely used as a fungicide, several microorganisms, including fungi, can degrade PCNB [6]. They can degrade PCNB into its derivatives, namely pentachloroaniline, pentachloro-thioanisole, pentachlorothiophenol, and pentachlorophenol (**Figure 3**). More likely, *Ganoderma* can also be one of the microorganisms that can degrade PCNB. This needs further studies.

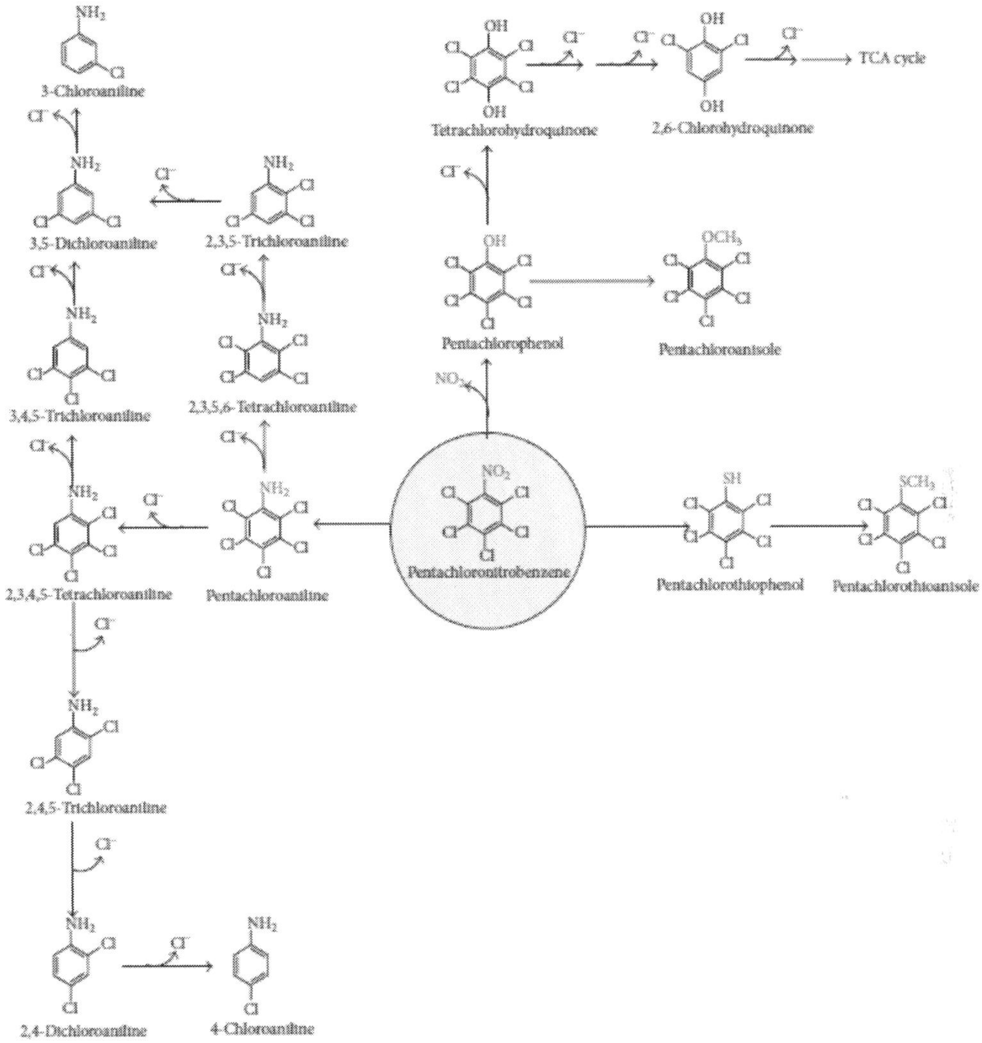

Figure 3. PCNB degradation into its derivatives via metabolism of microorganisms. Taken from Arora and Bae [6].

3.2 Guaiacol performance as substitute ingredients for PCNB

Ganoderma that grow on GSM+Guaiacol was somehow not showed a formation of brown halo which gives GSM its unique characteristic that can separate *Ganoderma* colony from others. Moreover, *Ganoderma* hyphae were not growing. Other fungi were also seen to be not growing either (**Figure 4**). Guaiacol is thought to be a substitute ingredient for PCNB due to its similarity in the chemical structure. Both of them are aromatic compound and *Ganoderma* has the enzyme that can utilize such aromatic compound [4]. However, the hypothesis was proved wrong based on the result of this study.

Ganoderma that grow on GSM+Guaiacol was somehow not showed a formation of brown halo which gives GSM its unique characteristic that can separate *Ganoderma* colony from others.

Moreover, *Ganoderma* hyphae are not growing. Other fungi were also seen to be not growing either (**Figure 4**).

Guaiacol has been used as an ingredient to distinguish whether the tested isolate has the ability to produce laccase, an extracellular enzyme which is grouped to oxidase enzymes and a part of ligninolytic system enzyme, by changing the colour of agar media into dark-brown [9]. Its mechanism is the same with tannic acid because both of guaiacol and tannic acid react to the presence of phenoloxidase [10] [9]. However, when they were combined, they did not change the colour of the media in this study. Guaiacol is thought to be a substitute ingredient for PCNB due to its similarity in the chemical structure. Both of them are aromatic compound and *Ganoderma* has the enzyme that can utilize such aromatic compound [11]. However, the hypothesis was proved wrong based on the result of this study.

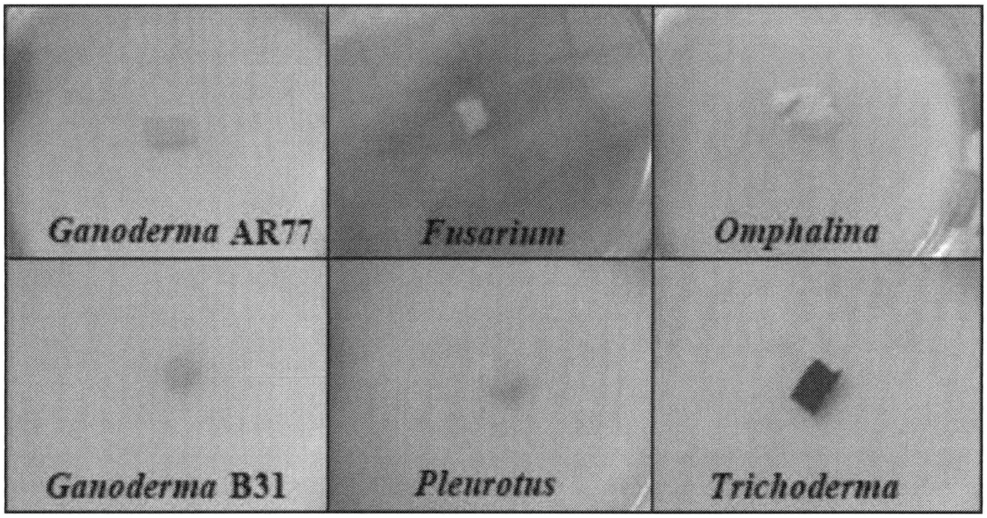

Figure 4. The morphology of the fungi that were grown on GSM added with guaiacol as substitute ingredients for PCNB. Noticed that GSM lose its unique characteristic and *Ganoderma* hyphae cannot grow either.

4. Conclusions

GSM could still form a brown halo when *Ganoderma* was inoculated on it even with the absence of PCNB and Ridomil. The brown halo was not observed when other fungi were grown on modified GSM. PCNB is thought to be the carbon source because the growth of *Ganoderma* is restricted without the presence of the compound in the medium. In spite of that, modified GSM can still be used as a medium that can selectively isolate *Ganoderma* from others when one's try to isolate *Ganoderma* whether from soil or oil palm tissue.

5. References

[1] Zafitra Z, Elfina Y and Ali M 2017 *Jurnal Online Mahasiswa (JOM) Bidang Pertanian* **4** 1
[2] Chang T 2003 *Plant Disease* **87** 1201
[3] Hasan Y and Turner P D 1998 *Planter* **74** 119
[4] Ariffin D, Idris A and Singh G 2000 *Ganoderma diseases of perennial crops* 49
[5] Muladi 1999 Government Regulation No. 85 /1999 on Toxic and Hazardous Waste Management. ed G Indonesia (Jakarta: Indonesian Ministry of the State Secretariat)
[6] Arora P K and Bae H 2014 *Journal of Chemistry* **2014** 12
[7] Naidu Y, Idris A S, Nusaibah S A, Norman K, Siddiqui Y and Mohammed C 2015 *Forest Pathology* **45** 474-83

[8] Cipriano-Salazar M, Rojas-Hernández S, Olivares-Pérez J, Jiménez-Guillén R, Cruz-Lagunas B, Camacho-Díaz L M and Ugbogu A E 2018 *Microbial Pathogenesis* **117** 255
[9] Xu C, Singh D, Dorgan K M, Zhang X and Chen S 2015 *Canadian Journal of Microbiology* **61** 745-52
[10] Sinsabaugh R L 2010 *Soil Biology and Biochemistry* **42** 391
[11] Surendran A, Siddiqui Y, Saud H M and Manickam N S A a S 2017 T *International Journal Of Agriculture and Biology* **19** 1437

IOP Conf. Series: Earth and Environmental Science **183** (2018) 012021 doi:10.1088/1755-1315/183/1/012021

The potency of an automated detection system for real-time pest monitoring from smartphones and spatial analysis for leaf-eating caterpillar attack evaluation

Henny Hendarjanti[1]*, S W Indratno[2], U T Ismunandar[1] and M H Sugeng W[1]

[1]PT Astra Agro Lestari Tbk. Jalan Pulo Ayang Raya Blok OR/I, Kawasan Industri Pulo Gadung, Jakarta Timur, Indonesia
[2]Bandung Institute of Technology, Statistics Research Group, Jalan Ganesha 10, Bandung, Indonesia

*Email: hennyhendarjanti@yahoo.com

Abstract. An early warning system in the oil palm cultivation is needed to know the early existence of pest quickly, precisely and accurately. The automated report is needed because manually or conventional methods of the pest existence in the field are not optimal and accurate. A web-based automatization is programmed for giving early information about pest and the potential natural enemies through daily report using the web which is integrated into the database. Moderate and severe pest attack will inform by short message services. So that, it is easier for management to know pest, natural enemies, and conservation beneficial plants. The data can also be used as a reference for controlling, evaluating, and making a decision as well as knowing real-time conditions exactly, quickly, and accurately. This technique is very important especially for controlling leaf-eating caterpillar and conserving its natural enemy's population in the field.

1. Introduction

The oil palm production has an important role for plantation and economic sector in Indonesia. One of the important problems in the oil palm plantation is pests which cause fruit bunch damage. The leaf-eating caterpillars eat oil palm leaf that causes a decline in the palm oil production. Reducing the amount and frequency of pesticide applications could not only help to reduce some of the unintended impacts of agricultural practices, but also could save time, labour and cost for the growers [1-3].

The early warning system is needed to minimize losses caused by the pest and to know the existence of natural enemies. This research applied automated detection system with web-based, Android, and spatial analysis. Non-stationary time series modelling, especially the family of autoregressive models used to predict the number of pests based on their historical data. We realized that there is some uniqueness of these pest data, i.e. the spike value that occurs almost periodically [4-6]. Through some simulations and case study, we obtained that the selection of constants factor has a significant influence on the model so that it could shoot the spikes value precisely.

2. Automated Early Warning System Processes

Mobile Early warning system was a daily application that has been used for pest and diseases observation. The application used gadgets based on Android with special design for PT Astra Agro Lestari Tbk. The data informed of the coordinate position of field condition actually through GPS. The recorded GPS data in the gadgets were transferred to the OTG USB device, after that the data were sent online and connected to the head office so that the data report was observed and analysed especially for the moderate and severe attacks, and directly communicate using SMS (short message services) [7].

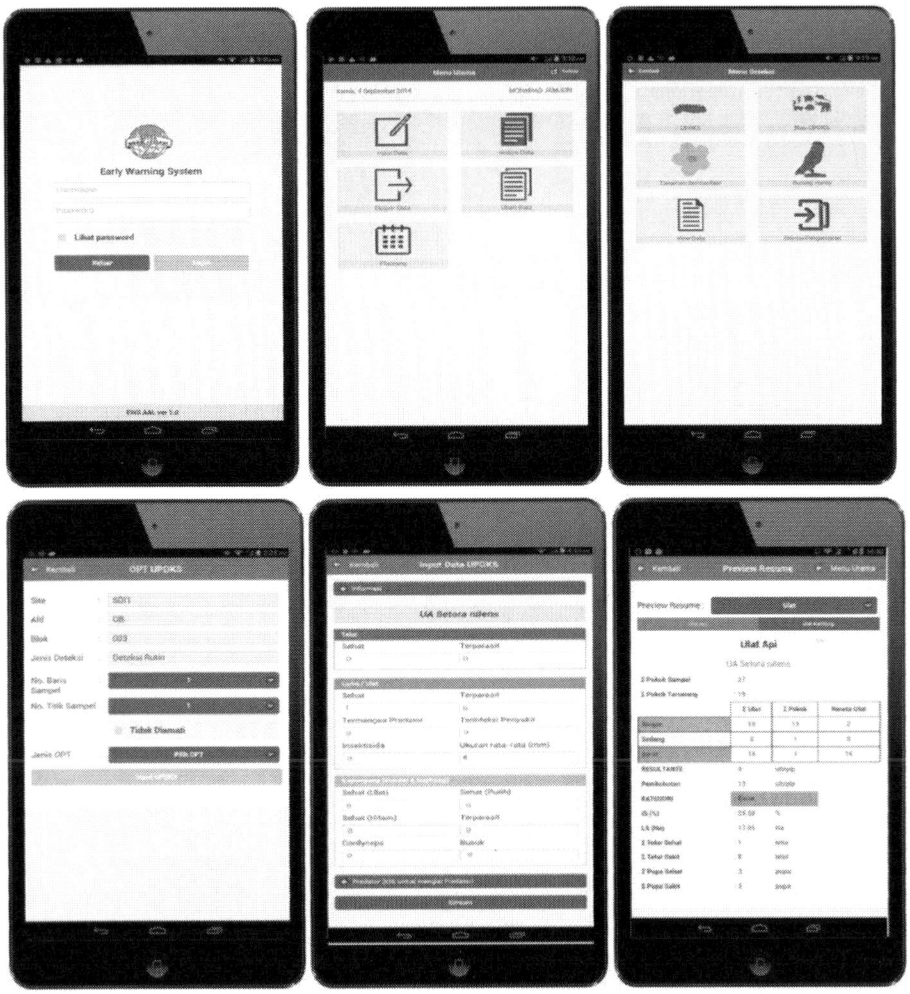

Figure 1. The mobile early warning system of Astra Agro Lestari, Tbk.

3. Automated Spatial Analysis

Through the automated early warning system, we conducted some statistical analysis related to the most recent field conditions. The automated system provides us with the following information, i.e: time series data, statistical descriptive data, and the spatial data of the pest as given in **Figure 2, 3, 4, 5, 6** and **7**.

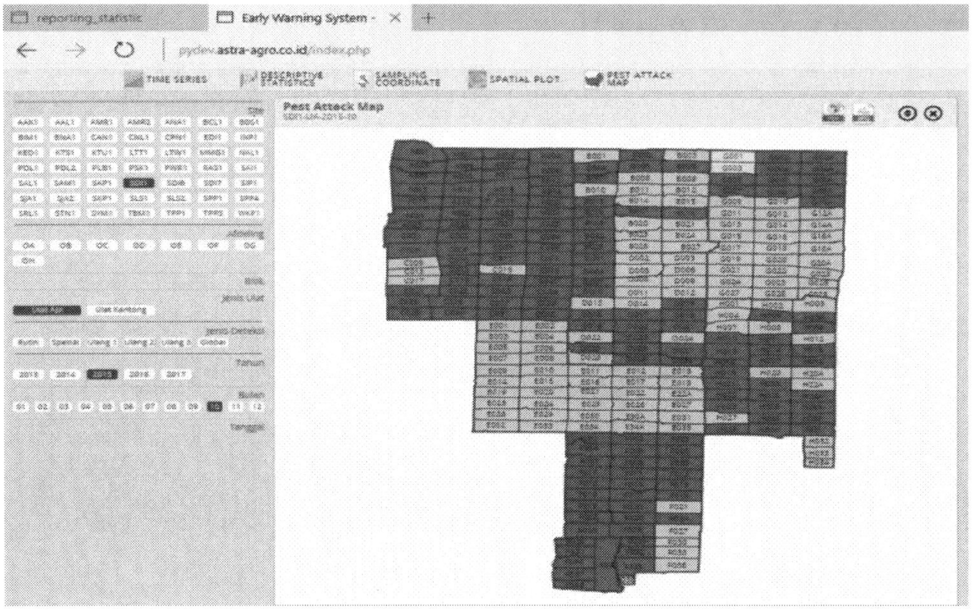

Figure 2. Areas of observation.

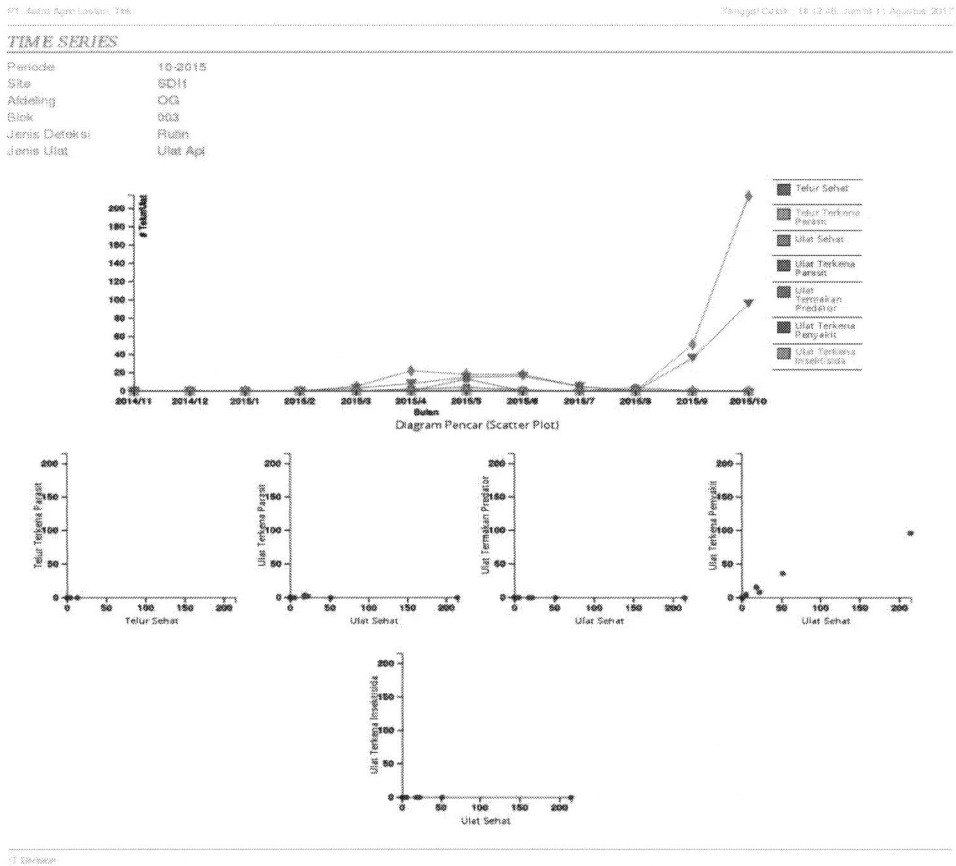

Figure 3. Time series data.

IOP Conf. Series: Earth and Environmental Science **183** (2018) 012021 doi:10.1088/1755-1315/183/1/012021

Figure 2 showed a visual representation of the automated early warning system. From the **Figure 2**, we chose an observation at each area, where the system provided the most recent reported data of the pest. From the reported data a map was made in the different colour which was red for severe, yellow for moderate, green for light, and grey for none.

The time series data help us in investigating the behaviour of pests based on time where we could check whether the pest has a pattern at particular months (**Figure 3**) [8]. Moreover, the system also gives a prediction of the number of pests based on the historical data. This prediction helps the decision maker to make a future plan related to preventing damages caused by the pest. In modelling the time series and prediction we applied the Zero-Inflated Poisson (ZIP), model. This model suitable for modelling time series of count where the occurrence of zero counts is higher than the non-zero count. In this model we assumed that the number of pests followed mass distribution function :

$$P(Y_t = k) = \begin{cases} \pi + (1-\pi)e^{-\lambda} & ,\text{for } k = 0 \\ \dfrac{(1-\pi)(\lambda^k e^{-\lambda})}{k!} & ,\text{for } k \neq 0 \end{cases},$$

where Y_t is the number of pests at a time t, π is the probability of extra zero and λ is the expected Poisson count. The previous information in the model based on the equation $\lambda_t = e^{\alpha_0 + \alpha_1 Y_{t-1}}$, where α_0 and α_1 are parameters which are estimated based on the historical count data [9, 10].

Descriptive Statistics

Periode	10-2015
Site	SDI1
Afdeling	OG
Blok	003
Jenis Deteksi	Rutin
Jenis Ulat	Ulat Api
Tanggal Pengamatan	23 Oktober 2015

Banyaknya Titik Sample 30

	Telur Sehat	Telur Terkena Parasit	Ulat Sehat	Ulat Terkena Parasit	Ulat Termakan Predator	Ulat Terkena Penyakit	Ulat Terkena Insektisida
#	0	0	214	0	0	96	0
Rataan	0	0	7.1333	0	0	3.2	0
Rataan bersyarat	-	-	11.8889 | 18	-	-	6.8571 | 14	-
Modus	0	0	0	0	0	0	0
Median	0	0	8	0	0	0	0
Minimum	0	0	0	0	0	0	0
Maximum	0	0	19	0	0	13	0
Jangkauan	0	0	19	0	0	13	0
Simpangan Baku	0	0	6.5692	0	0	3.9252	0
Variansi	0	0	43.154	0	0	15.4069	0
Skewness	0	0	0.1197	0	0	0.7839	0
Kurtosis	-3	-3	-1.4622	-3	-3	-0.5812	-3
Kuartil Ke-1	0	0	0	0	0	0	0
Kuartil Ke-2	0	0	8	0	0	0	0
Kuartil Ke-3	0	0	13	0	0	6	0

Figure 4. Descriptive statistic.

Another feature provided by the automated early warning system is the descriptive statistic given in **Figure 4**. Using this feature, we observed the basic statistics of the pest data, such as the mean, standard deviation, skewness, kurtosis, boxplot, and scatter plot. Descriptive statistics is a powerful tool in analyzing the distribution of the pest. Then we observed in a particular area how the pest data were spread. The outliers' data also helped us to check an unusual pattern (**Figure 4**). The boxplot was built based on the first (Q_1), second (Q_2) and third (Q_3) quantiles of the data. We labeled the data as outliers if the value was lower than the value of $\min(Q_1 - 1.5 \times (Q_3 - Q_1), X_{(1)})$, where $X_{(1)}$ is the

IOP Conf. Series: Earth and Environmental Science **183** (2018) 012021 doi:10.1088/1755-1315/183/1/012021

smallest value of the data, or the value is larger than the value of $\min\left(Q_3 + 1.5 \times (Q_3 - Q_1), X_{(n)}\right)$, where $X_{(n)}$ is the largest value of the data.

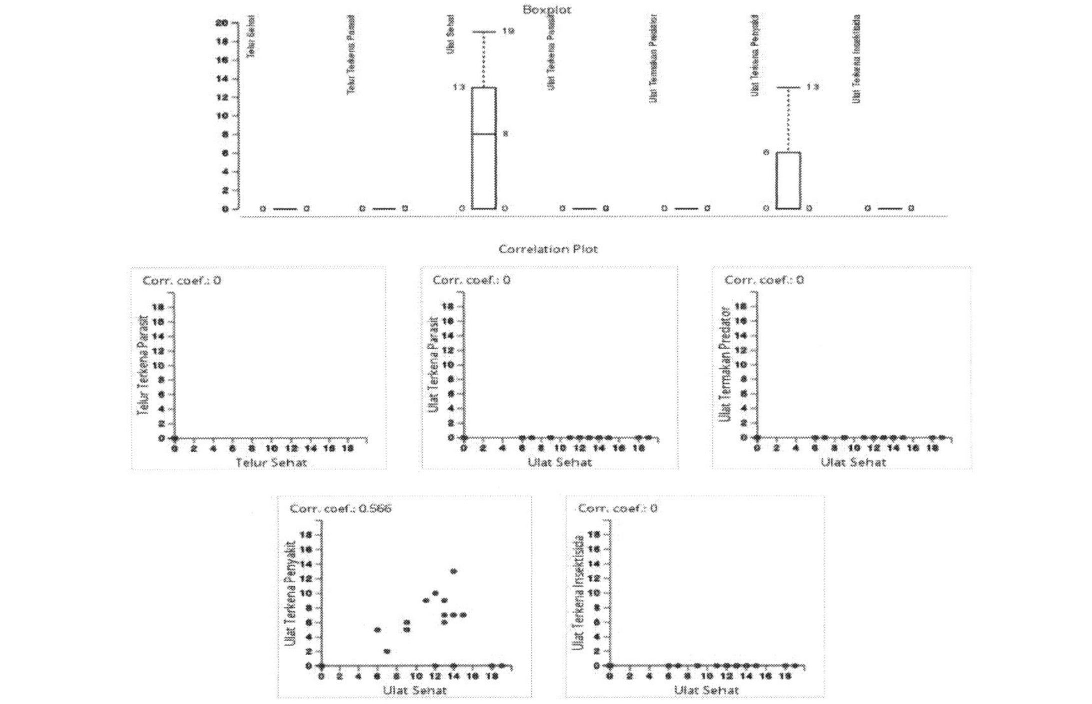

Figure 5. Boxplot and scatterplot.

Correlation between stages of the pest showed visually by scatter plots given in **Figure 5**. The correlation of stages of the pest was represented by the coefficient correlation. The closer this coefficient correlation to one, the stronger the linear relationship. But the small the coefficient correlation gave the weaker linear relationship between the two stages. The correlation between X and Y, $\rho_{X,Y}$, is based on the following formula:

$$\rho_{X,Y} = \frac{n\sum_{i=1}^{n} x_i y_i - \sum_{i=1}^{n} x_i \sum_{i=1}^{n} y_i}{\sqrt{\left(n\sum_{i=1}^{n} x_i^2 - \left(\sum_{i=1}^{n} x_i\right)^2\right)\left(n\sum_{i=1}^{n} y_i^2 - \left(\sum_{i=1}^{n} y_i\right)^2\right)}},$$

Where n is the number of pair X and Y, x_i is the i-th data of X and y_i is the i-th data of Y.

The third important feature was a spatial plot where we checked the sampling coordinates of the chosen area (**Figure 6**). This figure gave a number of the pest at each sample point. Based on these data the automated early warning system generated a distribution of the number of a pest at any site on the observed area (**Figure 7**). The colours represent the pest infestation levels. Therefore, the spatial plot used for identifying high-risk areas, while the spatial plot used for identifying high-risk areas from the pest outbreaks [7, 9].

IOP Conf. Series: Earth and Environmental Science **183** (2018) 012021 doi:10.1088/1755-1315/183/1/012021

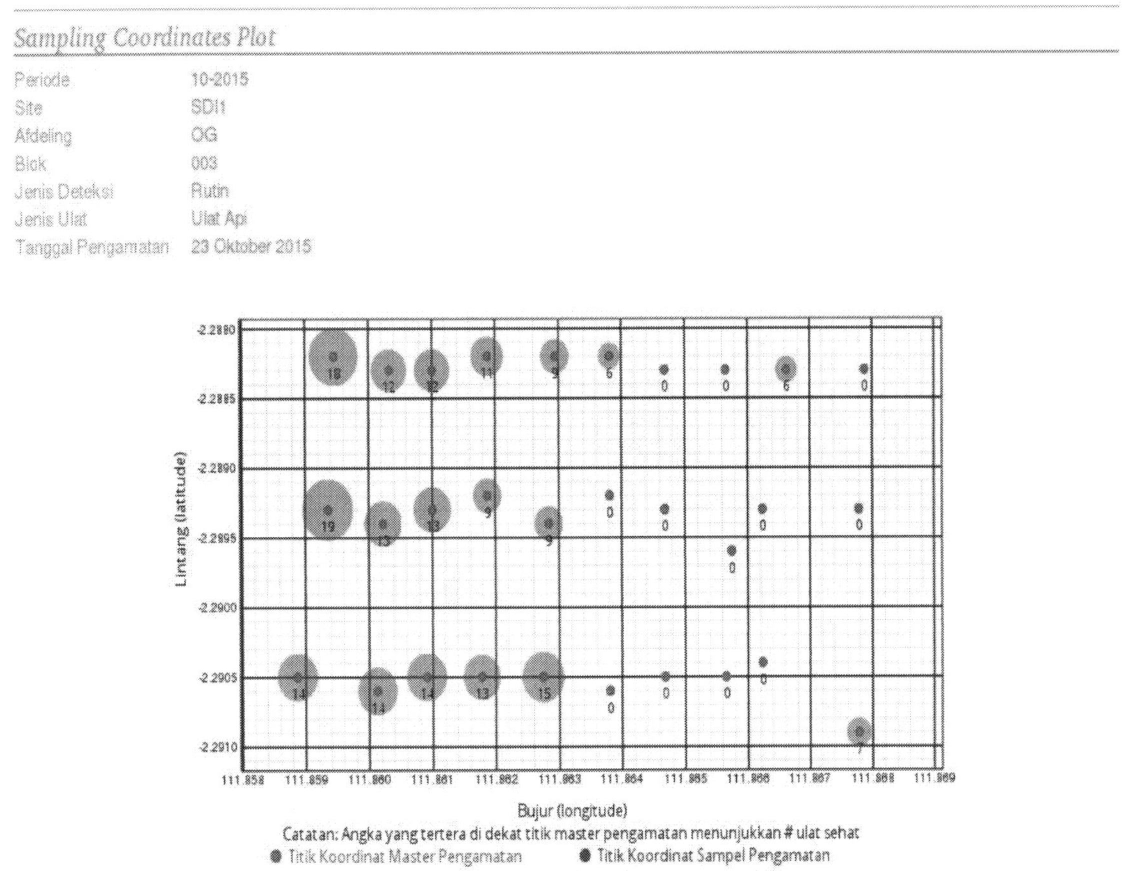

Figure 6. Sampling coordinates.

In the spatial analysis, we define $Y_{t,j}$ as the count at time t at the location j. In our case, there were about 30 sample locations and the following joint distribution density counted as function:

$$c(u_1, u_2, \dots, u_{30}) = \frac{e^{-z^T(\Sigma^{-1}-I)z}}{\sqrt{|\Sigma|}},$$

where $u_j = F(y_{t,j}), Z = \left(\Phi^{-1}(u_1), \Phi^{-1}(u_2), \dots, \Phi^{-1}(u_{30})\right)^T, \Sigma$ is the kernel matrix, $|\Sigma|$ is the determinant of the matrix Σ, and Φ^{-1} is the inverse cumulative distribution function of a standard normal. Based on the above joint distribution, the spread of the pest given in the spatial plot is predicted by a conditional probability [4].

4. Conclusions

The impacts of EWS automation are to facilitate observers of pest organisms in the oil palm plantations to fill sheets of student manual paper form softcopy that is stored and documented in the company's server; to inform direct field observation data on the same day (h+0), to alert decision-makers and staff *via* smartphone and email in case of pest attacks in moderate and severe categories, and to follow up controlling of plant-disturbing organisms immediately.

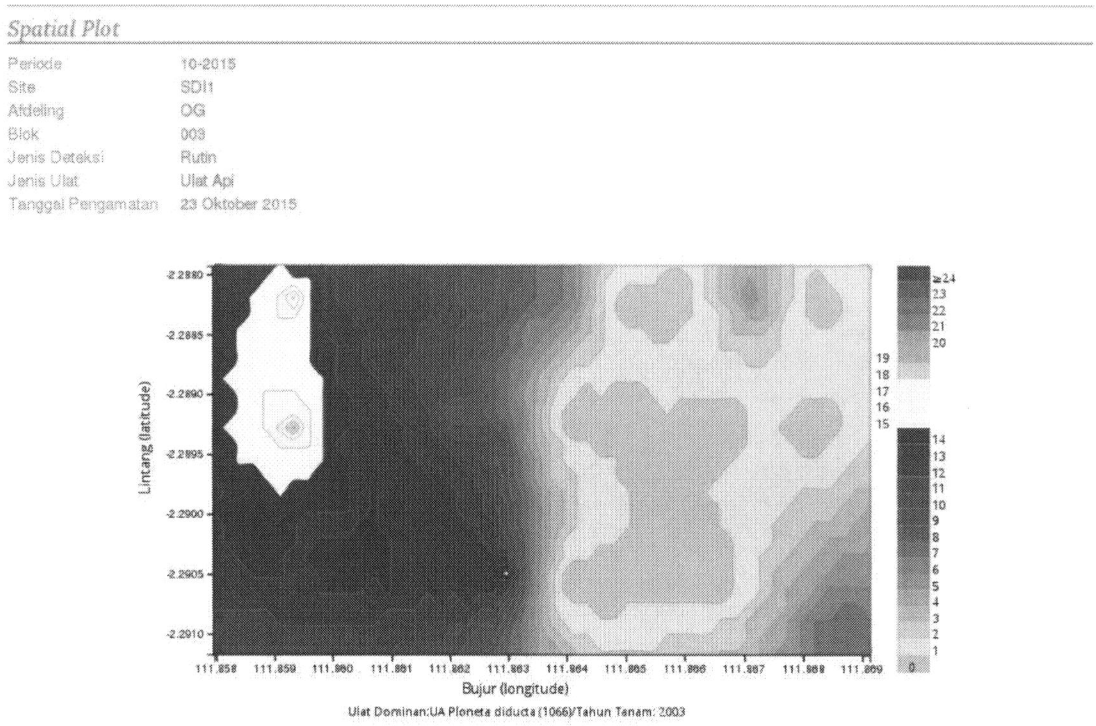

Figure 7. Spatial plot of the distribution of pest.

5. References

[1] Setyowati S, Nugraha R F and Mukhaiyar U 2015 *AIP Conf. Proc. (Bandung)* vol 1692 (New York: AIP Publishing) p 020011

[2] McMaugh T 2007 *Guidelines for surveillance for plant pests in Asia and the Pacific* (Canberra: ACIAR) No 119a 192p

[3] Cooke B J and Lorenzetti F 2006 *For. Ecol. Manage.* **226** 110

[4] Andrea S and Pasquale T 2014 *Plant Protect. Sci.* **50** 129

[5] Manyong V M, Legg C, Mwangi M, Nakato V, Coyne D, Sonder K, Bouwmeester H and Abele S 2008 *IV International Symposium on Banana: International Conference on Banana and Plantain in Africa: Harnessing International (Mombasa)* vol 879 (Arusha: AVRDC) pp 333

[6] Riggi L 2017 Integrated pest management across spatial scales [*Dissertation*] (Uppsala: Swedish University of Agricultural Science)

[7] Polukoskho S and Hofmanis J 2009 *Proc.7th Inter. Sci. Prac. Conf. (Volvograd)* vol 11 (Paris: Atlaltis Press)

[8] Cohen A L and Crowder D W 2017 *Current Opinion in Insect Sci.* **20** 13

[9] Deleon L, Brewer M J, Esquivel I L and Halcomb J 2017 *Crop Protec.* **101** 50

[10] Zuur A F, Savaliev A A and Leno E N 2012 *Zero-Inflated Models and Generalized Linear Mixed Models with R* (Newburgh: Highland Statistics Ltd.)

IOP Conf. Series: Earth and Environmental Science **183** (2018) 012022 doi:10.1088/1755-1315/183/1/012022

Effectiveness of Weed Solut-Ion as herbicide adjuvant to control weeds in oil palm plantations

Prabawati Hyunita Putri[1*] and D Guntoro[2]

[1]Pandawa Agri Indonesia, Jalan Teuku Cik Ditiro 1 No.11 Menteng, Jakarta 10350, Indonesia
[2]Department of Agronomy and Horticulture, Bogor Agricultural University, Kampus IPB Darmaga, Jl. Meranti, Babakan, Dramaga, Bogor 16680, Indonesia

*E-mail: prabawatihyunita@pandawaid.com

Abstract. The use of herbicides is an effective and efficient way to control weeds in oil palm plantations. However, overdoses application of herbicides is hazardous to the environment. Therefore, it is necessary to find a method to reduce hazardous effects caused by excessive doses of herbicides. Adjuvant herbicides are a well-known compound that could enhance herbicide performance. Weed Solut-ion (WS) contains 2-sodium amine allegedly reduces herbicide dosage at the same time enhances its performance. The objective of this research was to determine the response application of Weed Solution as herbicide adjuvant for paraquat and fluroxypyr by reducing the dosage so that the application can be more efficient and environmentally friendly. The experiment was conducted in the Socfin Group Plantation using a completely block design with two treatments (control; paraquat:WS = 50:50; fluroxypyr:WS = 50:50). The results showed that weeds control and regrowth rate from treatment (paraquat:WS = 50:50) were not significantly different from the control. The cost efficiency resulted from the treatment was 25.56%. The dose treatment of fluroxypyr combined with WS (doses 50:50) was not significantly different from control, and the cost efficiency was 36.43%.

1. Introduction

Weed is the one of a major problem in oil palm plantation because it will lead to a nutrient deficiency in oil palm tree that can cause yield loss [1]. Most of the weed control is using chemical or known as herbicides because of its effectiveness, less time consuming, and low cost compared with using mechanical control. Paraquat, fluroxypyr, metsulfuron, ammonium gluphosinate, and glyphosate are commonly used by planters to control weeds [2].

According to Aktar *et al.* [3], there are a lot of problems found because of the use of herbicides in uncontrolled doses which can lead to the damaging environment (contamination, kill non-target species etc.), weed resistance, and human health. Uncontrolled doses application happens because some cases showed that normal doses did not control weed so that planters using higher doses to get more effective in controlling weeds. This could lead to resistance and a higher amount of contamination in the environment.

Controlling the use and the doses of herbicides are needed to reduce their negative effect. Adjuvant herbicides are known as a compound that could enhance herbicides performances although the doses of herbicides decreased. Weed Solut-ion (WS) contains 2-sodium amine is alleged could reduce doses but enhance the performance of herbicides. The objective of this research was to know the response application of Weed Solution as herbicide adjuvant for paraquat and fluroxypyr by reducing the doses so that the application can be more efficient and environmentally friendly.

2. Materials and Methods

2.1. Place and Date

The experiment was conducted juvenile oil palm plantations: Matapao plantation Block 40 (date of planted in 2014; fluroxypyr's treatment) and Bangun Bandar plantation block 90 (date of planted 2014; fluroxypyr's treatment); block 31 (date of planted 2014; paraquat's treatment) from July to October 2015.

2.2. Treatment and Application

The area of the experiment was identified before application to know the population of weeds. The treatment of this experiment was using two active ingredients of herbicides i.e paraquat, fluroxypyr, and glyphosate. The dosage that used as a described in **Table 1** and **2**. Each of the treatments was mixed well before applied. The mixture was sprayed to all over the weed's surface until completely wet.

Table 1. Dosage treatment of paraquat, metsulfuron-methyl, and Weed Solut-ion (WS).

	Treatments	Paraquat (mL/Ha)	(WS)	Metsulfuron-methyl (g/Ha)
P1 (Control)	Paraquat	200	-	10
P2	Paraquat + WS	100	100	10
P3	Paraquat + WS	100	100	5

Table 2. Dosage treatment of fluroxypyr and Weed Solut-ion (WS).

	Treatments	Fluroxypyr (mL/Ha)	WS (mL/Ha)
F1 (control)	Fluroxypyr	40	-
F2	Fluroxypyr + WS	20	20

Table 3. Dosage treatment of glyphosate, 2,4 D, and Weed Solut-ion (WS).

	Treatment		
	Glyphosate (mL/Ha)	2,4-D (mL/Ha)	WS (mL/Ha)
G1 (control)	300	-	-
G2	150	-	150
G3	300	100	-
G4	150	50	200
G5	350	-	-
G6	175	-	175

2.3. Experimental design and observation

The experimental design was using a randomized complete block design with three replicates for each treatment. The layout of each treatment unit within a group is determined in such a way so that the target weed distributions are evenly the same. Post spraying assessments were carried out at weekly intervals on percentage weed killed and percentage weed regeneration (until 2 months). Variable of observation consisted of the efficacy of each treatment and cost efficiency. The effectivity of herbicides combined with Weed Solut-ion was compared with utilized doses (full doses, without WS). Analysis of variance was conducted and means were compared by Duncan's Multiple Range Test and Tukey at 5% level of significance.

IOP Conf. Series: Earth and Environmental Science **183** (2018) 012022 doi:10.1088/1755-1315/183/1/012022

3. Results and Discussion

3.1. The effectiveness of 2-Sodium Amine Combined with Fluroxypyr Treatment

The major weed population recorded in this experiment in Matapao and Kebun Bandar plantation was mainly *Mucuna bracteate*. This species supposed to be a land cover crop to enrich nutrient in the soil and decrease evaporation [4,5]. This species became a weed if uncontrolled because of its fast-growing and cover the immature oil palm tree quickly. **Table 4** and **Figure 1** showed successful weed control was achieved at 4 weeks after application.

Table 4. The effectiveness of fluroxypyr (F1, control) and the combination of fluroxypyr+WS (F2) at Matapao plantation.

Treatments	Replicates	Weed control (%)
	Fluroxypyr 40 mL/Ha	
	U1	100 a
F1	U2	100 a
	U3	100 a
	Fluroxypyr 20 mL/Ha+ 20 mL/Ha	
	U1	100 a
F2	U2	100 a
	U3	100 a

Figure 1. Effectiveness treatment of fluroxypyr combined with Weed Solut-ion (WS) at Matapao plantation. (A) control (F1) utilized doses before treatment; (B) F2 before treatment; (C) F1 at 4 weeks after application; (D) F2, combination of fluroxypyr and WS at 4 weeks after application.

Table 5 and **Figure 2** showed the successful effectiveness of fluroxypyr combined with WS compared with control at Kebun Bandar plantation. This result showed there is no significant difference between an application using normal or utilized doses of fluroxypyr and half doses of fluroxypyr combined with WS. All of *M. bracteata* were dried at 4 weeks after application. Fluroxypyr had systemic effects so that have successfully controlled the weeds started 2 weeks after treatment and completely dried at 1 month after treatment. *Mucuna bracteata* regeneration in both of treatment was faster than other weeds. They had regrowth at 4 weeks after treatment. According to Harahap *et al.* growth rate of *Mucuna bracteata* at the beginning of planting was 2-3 m^2 per month.

Table 5. The effectiveness of fluroxypyr (F1, control) and the combination of fluroxypyr+WS (F2) at Kebun Bandar plantation.

Treatments	Replicates	Weed control (%)
	Fluroxypyr 40 mL/Ha	
	U1	100 a
F1	U2	100 a
	U3	100 a
	Fluroxypyr 20 mL/Ha+ 20 mL/Ha	
	U1	100 a
F2	U2	100 a
	U3	100 a

Figure 2. Effectiveness treatment of fluroxypyr combined with Weed Solut-ion (WS) at Kebun Bandar plantation. (A) control (F1) utilized doses before treatment; (B) F2 before treatment; (C) F1 at 4 weeks after application; (D) F2, a combination of fluroxypyr and WS at 4 weeks after application.

Combination of fluroxypyr and Weed Solut-ion (WS) could kill weed although used a half dose of fluroxypyr. The active compound of WS enhances the performance of fluroxypyr to control *M. bracteata.* This adjuvant was allegedly acted as a surfactant to make penetration of fluroxypyr easier. *Mucuna bracteata* regeneration in both of treatment was faster than other weeds. They had regrowth at 4 weeks after treatment [6]. According to Harahap *et al.*[3] growth rate of *M. bracteata* at the beginning of planting was 2-3 m^2 per month [7].

3.2 Effectiveness of Weed Solut-ion Combined with Paraquat and Metsulfuron-methyl

Target weed in this experiment was *Stenochlaena palustris*. The result showed that percentage of weed control between normal doses (P1) and the combination of paraquat, metsulfuron-methyl, and Weed Solut-ion (WS) was not significantly different at 2 and 3 months after treatment (MAT) (**Table 6**). P2 treatment had the lowest percentage of dried weed than others, but the result was not significantly different. The result indicated that decreasing the recommended doses of herbicides (paraquat and metsulfuron-methyl) until 50% and combined with Weed solute-ion still effective to control *S.*

IOP Conf. Series: Earth and Environmental Science **183** (2018) 012022 doi:10.1088/1755-1315/183/1/012022

palustris. Regrowth rate of *S. palustris* at 3 months after the application was not significant differences between all treatments (**Figure 3**). P2 treatment showed a higher regrowth rate among other but not significant different with control (P1) and P3. This might be because the percentage of decrease population weed in this treatment was lower so that part of *S. palustris* had growth faster. According to Faiz [8], paraquat was more effective herbicides than glyphosate and ammonium glufosinate, but the chemical is not suitable for sustainability in oil palm. So that reducing the doses of paraquat could help planters to decrease the side effect of paraquat on the environment. The addition of Weed Solut-ion can reduce paraquat residue on oil palm weeding circle soil [9].

Table 6. The effectiveness of paraquat and combination of paraquat + methyl metsullfuron + WS at Matapao plantation.

Treatments	Paraquat (mL/Ha)	Metsulfuron-methyl (gr/Ha)	WS (mL/Ha)	Weed control (%)	
				2 MAT	3 MAT
P1	200	10	-	99.20 ± 2.65 a[1]	98.00 ± 1.06a
P2	100	10	100	91.67 ± 7.94 a	90.60 ± 6.24a
P3	100	5	100	98.67 ± 0.23 a	97.87 ± 0.57a

[1]Means followed by the same letter are not significantly differenced at 5% level using DMRT

Figure 3. Effectiveness treatment of paraquat combined with Weed Solut-ion (WS) at Matapao plantation. (A) control (P1) utilized doses before treatment; (B) P2 before treatment; (C) P3 before treatment; (D) P1 at 2 months after treatment (MAT); (E) P2 at 2 MAT; (F) P3 at 2 MAT; (G) P1 at 3 MAT; (H) P2 at 3 MAT; (I) P3 at 3 MAT.

3.3 Effectiveness of Weed Solut-ion Combined with Glyphosate and 2,4 D

Analysis of weed population in block 36 Matapao plantation before the application was dominated by weeds as shown in **Table 7** and **Figure 4.** Almost 75% of vegetation was dominated by *Ottochloa nodosa* (**Figure 4**).

Table 7. Analysis vegetation of weed in block 36 Matapao plantation.

Weeds	Weed population (%)
Ottochloa nodosa	75
Cynodon dactylon	10
Axonopus compressus	5
Ageratum conyzoides	3
Synedrella nudiflora	2
others	5

Figure 4. Vegetation of weeds in block 36 Matapao plantation. (A) *Ottochloa nodosa*; (B) *Cynodon dacttylon*; (C) *Ageratum conyzoides*; (D) *Axonopus compressus*; and (E) *Synedrella nudiflora.*

Table 8. The effectiveness of glyphosate, 2,4 D, and combination with Weed Solut-ion (WS)

Treatments	Glyphosate (mL/Ha)	2,4 D (mL/Ha)	WS (mL/Ha)	% Weed control	% Weed regrowth
G1	300	-	-	100.00 a[1]	4.00 ab
G2	150	-	150	99.27 ab	4.33 ab
G3	300	100	-	99.87 a	4.00 ab
G4	150	50	200	99.00 ab	6.60 a
G5	350	-	-	99.87 a	2.80 b
G6	175	-	175	98.33 b	5.18 ab

[1] Means followed by the same letter are not significantly differenced at 5% level using DMRT

The result showed that treatment all treatment was not significantly different than control (**Table 8** and **Figure 5**). These indicated that decreasing the recommended doses of herbicides and combined with Weed Solut-ion still effective to control weeds. Glyphosate had systemic effects so that have successfully controlled the weeds starting 3 weeks after treatment and completely dried at 1 month after treatment. The weeds initially became yellowish at 2 weeks after treatment and turned brown or dried then finally dried up at 1 month after treatment. The regrowth rate of weeds started at 2 months after treatment, all treatment was not significantly different with control in growth rate (**Figure 6**).

Figure 5. The effectiveness of glyphosate, 2,4 D, and combination using Weed Solut-ion (WS) at 1 month after treatment. (A) G1, 300 mL/Ha glyphosate; (B) G2, 150 mL/Ha glyphosate + 150 mL/Ha WS; (C) G3, 300 mL/Ha glyphosate + 100 mL/Ha 2,4 D; (D) G4, 150 mL/Ha glyphosate + 50 mL/Ha 2,4 D + 200 mL/Ha WS; (E) G5, 350 mL/Ha glyphosate; (F) G6, 175 mL/Ha glyphosate + 175 mL/Ha W.

Figure 6. Regrowth of glyphosate, 2,4 D, and combination using Weed Solut-ion (WS) at 2 months after treatment. (A) G1, 300 mL/Ha glyphosate; (B) G2, 150 mL/Ha glyphosate + 150 mL/Ha WS; (C) G3, 300 mL/Ha glyphosate + 100 mL/Ha 2,4 D; (D) G4, 150 mL/Ha glyphosate + 50 mL/Ha 2,4 D + 200 mL/Ha WS; (E) G5, 350 mL/Ha glyphosate; (F) G6, 175 mL/Ha glyphosate + 175 mL/Ha WS.

3.4 Cost Efficiency

Cost comparison showed recommendation dosages of herbicides to be more expensive than combination herbicides with Weed Solut-ion (WS) (the price of herbicides and Weed Solut-ion was followed the price list in Socfin group plantation in 2015). The cost/Ha of fluroxypyr was IDR 6,275 and cost/Ha of combination Fluroxypyr and WS was IDR 3,897, so the difference was about IDR 2,378. In one year, plantation could save up to IDR 28,536/Ha/year and the cost efficiency was up to 37.90%. The cost/Ha of paraquat and metsulfuron-methyl (P1) was IDR 14,112, for P3 cost/Ha, was IDR 10,856 and the cost gap was about IDR 3,265. In one year, plantation did twice rotation and cost efficiency for 35,000 Ha was up to 23.07%. cost/Ha of glyphosate and 2,4 D was about IDR 15,721, and its combination with WS was about IDR 13,561 so that the efficiency was about 13.74%.

4. Conclusions

Application of 20 mL/Ha fluroxypyr and 20 mL/ha Weed Solut-ion (WS) was efficient to control weed in immature oil palm tree. Doses treatment used paraquat 100 mL/Ha, 5 g/Ha, and 100 mL/Ha WS had the same effectiveness with utilized doses. Combination of 150 mL/Ha glyphosate and 150 mL/Ha WS could control weed growth. Therefore, it is more economical to use a combination of Weed Solut-ion and a half doses of herbicides.

5. References

[1] Ofusu-Budu KG *et al.* 2014. Effect of glyphosate on weed control and growth of oil palm at an immature stage in Ghana. *International Journal of Agronomy and Agricultural Research.* **4** 1
[2] Latiff MR, Arbain MA, and Tey CC 2009 *Proc. Int. Palm Oil Cong.* (Kuala Lumpur: MPOC) **1** 1
[3] Aktar MW, Sengupta D and Chowdhury 2009 *Intedisc toxicol.* **2** 1
[4] Goh YT *et al.* 2016 *Oil palm bulletin.* **68** 16
[5] Laksono P B, Wachjar A and Supijatno 2016 *Jurnal Agronomi Indonesia.* **44** 104
[6] Kuntom A. *et al.* 2007 *Oil palm bulletin.* **54** 52
[7] Harahap IY, Taufiq CH and Simangunsong G 2008 *Mucuna bracteate* (Medan: Pusat Penelitian Kelapa Sawit)
[8] Faiz A 1991 *Proc. 13th Asian Pacific Weed Sci. Soc. Conf.* (Jakarta: APWSS) **1** p127
[9] Guntoro D, Lontoh A P and Putri N H 2016 *Proc. Seminar Nasional Perhorti dan Peragi* (Makassar: Perhorti) 402 (*In Indonesian)*

IOP Conf. Series: Earth and Environmental Science **183** (2018) 012023 doi:10.1088/1755-1315/183/1/012023

Inhibitory effect of phenolic acid on *Ganoderma boninense* enzyme as an approach on *Ganoderma* infection

Irma Kresnawaty[1]*, D D Eris[1], A S Mulyatni[1] and H T Prakoso[1]

[1]Indonesian Research Institute for Biotechnology and Bioindustry, Jalan Taman Kencana No.1, Bogor 16128, Indonesia

*E-mail: irma.kresnawaty@iribb.org

Abstract. *Ganoderma boninense,* the causal of basal stem rot (BSR) has already caused a major loss of revenue in Indonesia oil palm industry. Many methods were used to control *G. boninense* attacks. Phenolic compounds could be accumulated when *G. boninense* attacks oil palm plants and these compounds exhibit anti-*Ganoderma* activity. The aim of this research was to reveal the inhibitory capacity of phenolic acid against ligninolytic enzyme activity of *G. boninense*. Darkish-red colour around the colony of *Ganoderma* on lignin selective medium was observed. Extraction of crude enzymes resulted in 628.49 U/mL Laccase enzymes, 360.42 U/mL MnP enzymes, and 632.62 U/mL LiP enzyme from three replicates. We indicate that the protein concentration of the sample between 30-60 ppm according to standard series. By purified the enzyme three fractionation were obtained consist of fraction 5 which had 1254.4; 15.5; 548.39 U/mL, fraction 6 had 1394.21; 294.77; 141.58 U/mL, and fraction 7 had 196.53; 20.66 and 161.29 U/mL for the activity of laccase, MnP and LiP enzyme respectively. The addition of inhibitor resulted in the decrease of ligninolytic enzyme activity. Both syringic and caffeic acid showed a significant inhibition effect to *Ganoderma*.

1. Introduction

Basal stem rot (BSR) disease caused by *Ganoderma boninense* has already brought major loss of revenue in Indonesian oil palm industry. Currently, *G. boninense* can be found in almost in all oil palm plantations in Indonesia, although the incidence of the disease varies. The disease progresses rapidly not only in the mineral soil but also in peat lands. The incidence of *Ganoderma* disease reported most on nutrient-poor soils. *G. boninense* substrates in oil palm plantations are very abundant and alternative host disease is very broad [1]. Many methods are used to control *G. boninense* attacks, including clean and clearing the field method, crop rotation, eradication of infected plants, the usage of fungicides and biological control, and disease resistance improvement [2].

Chong *et al.* [3] reported that phenolic compounds accumulated when *G. boninense* attacked oil palm plants and these compounds exhibited anti-*G. boninense* activity. A group of phenolic acid has been reported as an enzyme inhibitor. Srivastava *et al* found that 10 phenolic acids from plants inhibited mycelium growth and its ligninolytic and pectinolytic enzymes activities [4]. Chong *et al* showed that phenolic acid especially syringic acid could inhibit the mycelium growth of *G. boninense in vitro* at a certain concentration [3]. *G. boninense* enzyme activities from corresponds to lignin and all other major structural cell wall polymers were detected together with their effect on host wall composition during *G. boninense* infection [5]. There is a high possibility that the phenolic compounds could interrupt *G. boninense* enzymes via competitive inhibitor mechanisms. Meanwhile, Widiastuti *et al* [6] funded by the State Ministry of Research and Technology are developing organic compound formula as bio-fungicide to *G. boninense*. In the same year, Eris *et al* [1] funded by IBRIEC itself is working on the induction of phenolic acid in oil palm by incorporating chitosan into the planting medium in the nursery stage. For supporting the research that mentioned later, information on the inhibitory effect of phenolic acid to *G. boninense* ligninolytic enzyme is needed. Therefore, the aim of this research is to see whether phenolic acid could inhibit ligninolytic enzyme activity of *G. boninense* or not.

2. Materials and Methods

2.1 Solid state fermentation

An unproductive 15-year-old oil palm tree was cut off, shredded and ground to obtain its powder form. Oil palm powder was then weighed to 20 grams, stored in jam bottle, and autoclaved. Several pieces of *G. boninense* inoculum in PDA were inoculated into the substrate (oil palm powder in the jam bottle). In addition, 2 mL of sterile water was added in order to make the condition moistly favourable for the fungus growth. The incubation period was 30 days.

2.2 Crude ligninolytic enzymes extraction

After incubated for 30 days, 20 mL of 0.2 M pH 7 phosphate buffers were mixed to the inoculated oil palm powder and the mixture was ground using mortar. Crude Ligninolytic enzymes were extracted by filtering the liquids resulted from grinding using Whatman paper No. 41.

2.3 Ligninolytic enzyme activities

Laccase activity was measured by spectrophotometry, as ABTS oxidation at 420 nm ($^{TM}{}_{420}$ = 36,000 M^{-1} cm^{-1}). The reactive mixture contained per 1 mL: 0.6 mL acetate buffer 0.5 M pH 5, 0.1 mL ABTS solution 1 mM and 0.3 mL enzyme filtrate.

Manganese peroxidase activity was also measured by spectrophotometry and the amount resulted from detraction of enzyme activity measured with and without $MnSO_4$ at 465 nm ($^{TM}{}_{465}$ = 121,000 M^{-1} cm^{-1}). The first reactive mixture contained 0.1 mL sodium lactate buffer 50 mM pH 5, 0.1 mL guaiacol 4mM, 0.2 mL $MnSO_4$ 1 mM, 0.1 mL H_2O_2 1 mM, 0.3 mL distilled water and 0.2 mL enzyme filtrate. The second reactive mixture was the same as the first one but without addition $MnSO_4$ thus to make the volume 1 mL, distilled water volume was added to its volume becomes 0.5 mL.

Lignin peroxidase activity was evaluated by UV spectrometry of the veratryl aldehyde produced ($^{TM}{}_{310}$ = 9300 M^{-1} cm^{-1}) during veratryl alcohol oxidation at 310 nm. The reactive mixture contained 0.2 mL acetate buffer 0.05 M pH 5, 0.1 mL veratryl alcohol 8 mM, 0.05 mL H_2O_2 1 mM, 0.45 mL distilled water, and 0.2 mL enzyme filtrate.

The absorbance of all enzyme assays was read at 0 and 30 minutes. For all enzymes under evaluation, one activity unit was defined as the amount of enzyme necessary to oxidize 1 nmol of substrate per minute. The specific activity was expressed as units per milligram of protein.

2.4 Addition of a hypothesized inhibitor

Syringic acid and caffeic acid solution were made which each of both concentrations were 0, 25, 50, and 100 ppm. These substances were examined to see whether they have an inhibition effect or not by adding 1 mL of each concentration of both substances to the enzyme reaction mixture and subsequently, the enzyme activities were checked using the aforementioned methods.

2.5 Purification method

Crude enzymes were purified using a gel permeation chromatography method. Sephadex G-100 and phosphate buffer were used as stationary and mobile phase, respectively. Ten grams of Sephadex G-100 was suspended with 300 mL phosphate buffer 0.1 M pH 6. Column packaging was prepared through stoppage of the column in the inside using cotton that drabble with the eluent to be used. The homogenized stationary phase was slowly poured against the column's wall using a pipette and allowed to stable until it was separated into two phases. The column flow rate was arranged and calculated (drop/min) so that the flow was moderate. Enzyme extract was then added until it reached the boundary of the stationary phase and the eluent was slowly added through the column's wall using a pipette, consecutively. Every 10 mL of liquid dripping from the column was collected for measurement of the total amount of protein using spectrophotometer at 280 nm. After the enzyme was purified, ligninolytic enzymes activities were measured once again with and without the hypothesized inhibitors.

3. Results and Discussion

3.1. The activity of Ganoderma boninense *enzyme*

Palm trunk powder extracted with 0.2 M phosphate buffer pH 7. Extraction was done with pH 7 phosphate buffer to maintain the pH of the protein that requires a certain pH range and prevents damage to the structure of the protein. Meanwhile, according to Bannerman *et al.* in Wulaningtyas [7], ligninolytic enzymes were extracted with phosphate buffer pH 7 because white root fungi (WRF) generally live at neutral pH. Otherwise, the structure of the protein will be denatured. Extraction was carried out to obtain crude enzyme that would be used in the enzyme activity assay.

Ligninolytic enzymes were on of the extracellular enzymes because of their role in the degradation process. So, it can be detected by using crude enzyme obtained from the supernatant (**Figure 1**). The main function of the extracellular enzyme is necessary to carry out nutrients around it, allowing the nutrients to enter the cell and so it can be absorbed directly by fungal hyphae [7].

WRF is a fungus that is capable of destroying all the major constituent components of wood (cellulose, hemicellulose, and lignin), thus allowing to reduce the content of xylem and phloem as the main constituent of wood logs and cause to become brittle. The process is aided by the enzyme ligninolytic which include laccase and peroxidase enzyme (MnP and LiP) [8]. *Ganoderma boninense* enzyme activity in this study was much higher compared with WRF *Omphalina* sp. and *P. ostreatus* with an average value of laccase, MnP and LiP consecutive 682.49 U / mL, 360.42 U / mL, and 632.62 U / mL. The difference in result might be caused by the different of likelihood, the difference of isolates and substrate used as well.

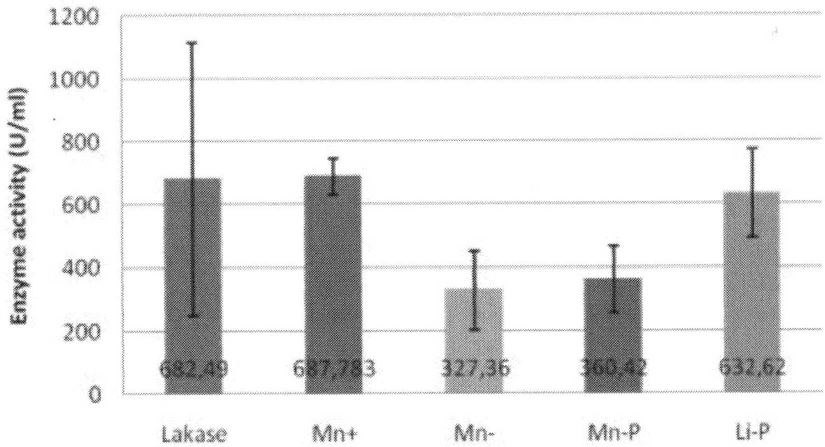

Figure 1. The ligninolytic enzyme activity of *Ganoderma boninense.*

3.2. Purification of an enzyme of Ganoderma boninense

Manganese peroxidase is an enzyme containing heme and is generally considered to be essential for lignin degradation in vivo. Manganese peroxidase catalytic cycle similar to lignin peroxidase, with the addition of Mn^{2+}, were oxidized to Mn^{3+} during the reaction. One unit of MnP proportional to 1 nmol of product produced per minute. Differences MnP considerable value between sample code 11, 17, and 25, probably due to the rate of growth of mycelium different in each bottle an hour so that the resulting value of different activities.

Lignin peroxidase is also a heme-containing peroxidase. Not all JPP produce lignin peroxidase. But on JPP *G. boninense* produce all the enzymes lignolytic even with a fairly high value when compared with other JPP. The highest yield was obtained from the sample code number 25 in the amount of 974.91 U / mL followed by sample code numbers 17 and 11 of 462.37 U / mL and 460.57 U / mL. The enzyme activity was measured at minute 0 and 30, the difference in absorbance at the time shall be included in the formula for calculating enzyme activity. Enzyme activity graph the relationship between the sample code with the activity of the enzyme produced. Graph enzyme activity is presented in **Figure 1**.

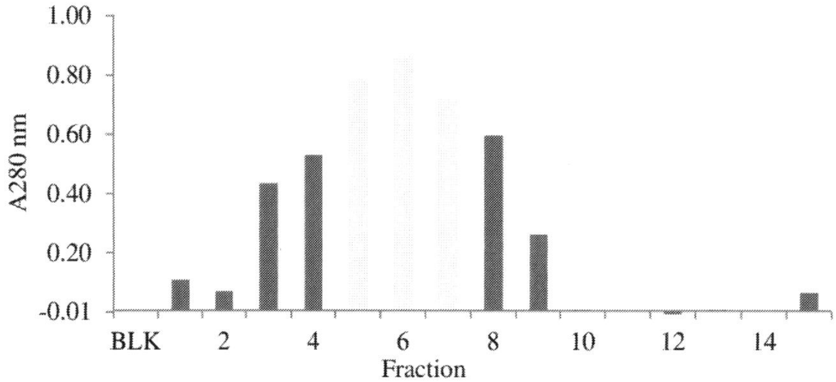

Figure 2. The absorbance of the eluate from gel permeation chromatography for partial purification.

To separate each ligninolytic enzyme, gel chromatography was conducted whereas the separation based on the different MW. We collected 15 fractions of eluate which were fraction 5,6 and 7 had the highest protein concentration (**Figure 2**). Based on of the MW, theoretically laccase which has the highest MW (50-110 kDa) will be collected first, then MnP (46 kDa) and LiP (38-46 kDa) [11]. From the three fraction, we analyzed the enzyme activity. Fraction 5 had 1254.4; 15.5 and 548.39 U/mL for laccase, MnP and LiP respectively. Meanwhile fraction 6 had 1394.21; 294.77; and 141.58 U/mL, and fraction 7 had 196.53; 20.66 and 161.29 U/mL for laccase, MnP and LiP respectively. Aslam and Asgher also found Laccase and MnP enzyme in the same fraction [9].

Because of the wide range of laccase MW, this enzyme was still found in later fraction be in conformity with the smaller one (MnP and LiP). It was too difficult to separate MnP and LiP because of the similar MW. *Ganoderma* has the highest activity of Laccase among the other lignolytic enzyme. Laccase is the oxidase enzyme which caused the injured area and discolours of the plant [10]. Fungal laccase plays a role in pigment formation in spores, detoxification of phenol compound produced during lignin degradation and acts synergistically with peroxidases and another enzyme in the breakdown of lignin [11]. Laccase can oxide a range of compounds such as diphenol, aryl diamine and aminophenol [12]. It assumed virulence of *G. boninense* is also determine with its high laccase activity.

3.3. Addition of Inhibitor

Effect of the inhibitor on ligninolytic enzymes needs to be investigated in vitro as possible control chemicals of *Ganoderma* on oil palm. Syringic and Caffeic acid have a similar structure with a substrate of the ligninolytic enzyme. Lignin consists of phenyl propanoic (p-hydroxyphenyl (H), guaiacyl (G) and syringyl (S) unit). The addition of inhibitor brought the decrease of laccase activity.

Syringic acid gives higher inhibition effect than caffeic acid. It could be caused by palm oil as woody angiosperms (hardwoods) consists of S that was the predominant phenolic component (important because lignin is a polyphenol), which comprised 65.6–68.5% of the total phenolic monomers in the oxidation mixtures. The increase of inhibitor brought the high inhibitory effect. Syringic has OH group that could assume has the ability to trap hydrogen peroxide derived hydroxyl radicals (OH•) and may be useful for inhibiting lignin peroxidase enzymes e.g. manganese peroxidase (MnP), lignin peroxidase (LiP), and laccase as found in benzoate [13].

Laccase oxidizes various phenolic compounds including lignin through phenoxy radical formation of phenoxy ring. Laccase using O_2 as oxidant and can oxidize phenolic compounds on a non-specific condition. Laccase enzyme activity obtained from the experiment were 188 U / mL, 115.05 U / mL, and 1743.98 U / mL for sample 11, 17, and 25 respectively (**Figure 3**). Sample 25 produced the highest laccase enzyme activity, possibly due JPP inoculated grow well so that the enzyme can degrade lignin in it.

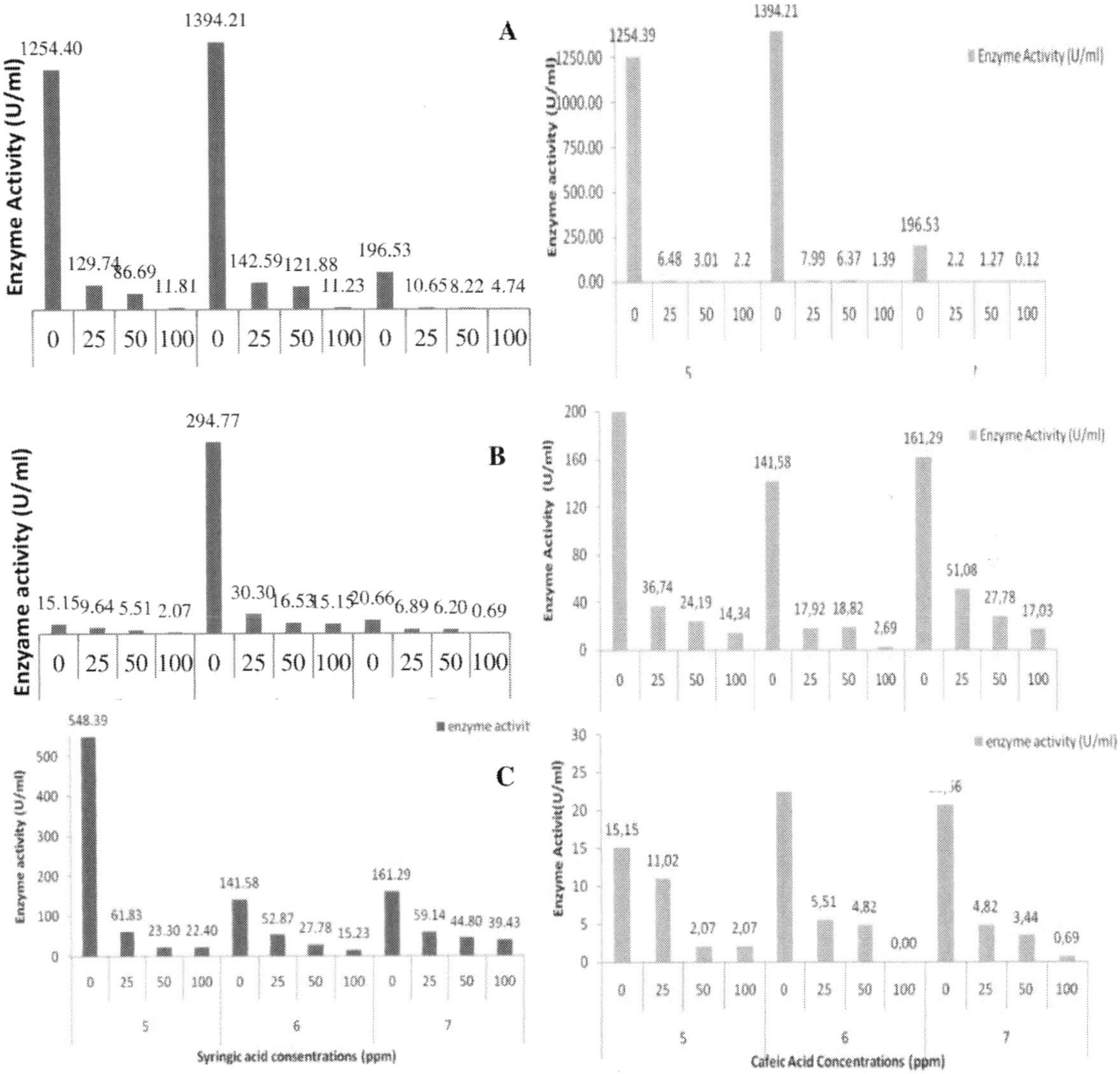

Figure 3. Changes in Laccase (A), MnP (B) and LiP (C) enzyme activity after several concentration additions of Syringic and Cafeic acid.

IOP Conf. Series: Earth and Environmental Science **183** (2018) 012023 doi:10.1088/1755-1315/183/1/012023

4. Conclusions

Extraction of crude enzymes resulted in 628.49 U/mL Laccase enzyme, 360.42 U/mL MnP enzyme, and 632.62 U/mL LiP enzyme from three replicates. We indicate that the protein concentration of the sample between 30-60 ppm according to standard series. By purified the enzyme three fractionation were obtained consist of fraction 5 which had 1254.4; 15.5; 548.39 U/mL, fraction 6 had 1394.21; 294.77; 141.58 U/mL, and fraction 7 had 196.53; 20.66 and 161.29 U/mL for the activity of laccase, MnP and LiP enzyme respectively. The addition of inhibitor resulted in the decrease of ligninolytic enzyme activity.

5. References

[1] Eris D D, Prakoso H T, Saragih A H and Darmono T W 2012 *Menara Perkebunan* 2012 **80** 8
[2] Sariah M and Zakaria K 2000 The use of soil amendments for the control of basal stem rot of oil palm seedlings In: Flood J, Bridge P, Holderness M, eds. *Ganoderma Diseases of Perennial Crops* (United Kingdom: CABI Publishing)
[3] Chong K P, Atong M and Rosall S 2012 *Plant Pathol* **10** 1
[4] Srivastava P, Ardersen PC, Marois J J, Wright D L, Srivastava M and Harmon P F 2013 *Crop Protection* **43** 146
[5] Cooper R M, J Flood, R W Rees 2011 *The Planter* **87** 515
[6] Widiastuti H, Santoso D, Putra S M, Wiramihardja M, Farida A, Marahimin B, Panjaitan K and Sinaga J 2013 *Menara Perkebunan* **81** 41
[7] Wulaningtyas A 2009 Aktivitas Enzim Lignolitik Jamur Pelapuk Putih *Omphalina* sp. dan *Pleurotus ostreatus* pada Medium Tandan Kosong Kelapa Sawit (TKKS) [Thesis] Bogor: FMIPA UNB
[8] Howard L, Abotsi E. van R J and Howard S 2003 *African J Biotechnol* **2** 602
[9] Aslam S and Asgher M 2011 *African Journal of Biotechnology* **10** 17875
[10] Roozbeh H, Yusof NA and Dutse SW 2013 Detection and control of *Ganoderma boninense*: strategies and perspectives. Springer Plus 2:555 http://www.springerplus.com/content/2/1/555
[11] Eggert C, Temp U and Eriksson K E 1997 *Febs Letters* **407** 89
[12] Xu F 1996 *Biochem* **35** 7608
[13] Paterson R R, Meon S, Abidin M Z and Lima N 2008 *Current Enzyme Inhibition* **4** 172